Jan Nesemann

PT-Symmetric Schrödinger Operators with Unbounded Potentials

VIEWEG+TEUBNER RESEARCH

Jan Nesemann

PT-Symmetric Schrödinger Operators with Unbounded Potentials

VIEWEG+TEUBNER RESEARCH

Bibliographic information published by the Deutsche Nationalbibliothek
The Deutsche Nationalbibliothek lists this publication in the Deutsche Nationalbibliografie;
detailed bibliographic data are available in the Internet at http://dnb.d-nb.de.

Dissertation Universität Bern, 2010

1st Edition 2011

Editorial Office: Ute Wrasmann | Anita Wilke

Vieweg+Teubner Verlag is a brand of Springer Fachmedien.
Springer Fachmedien is part of Springer Science+Business Media.
www.viewegteubner.de

Cover design: KünkelLopka Medienentwicklung, Heidelberg
Printed on acid-free paper
Printed in Germany

ISBN 978-3-8348-1762-4

Acknowledgment

I would like to express my gratitude to my supervisor Prof. Dr. C. Tretter for giving me the opportunity to write this thesis and for helping and supporting me during the past years. In particular, I would like to thank her for providing my wife and me with the great chance to come here to Switzerland. This was a welcome change which put us in the position to strike new paths.

Also, I would like to thank the members of the Applied Analysis Group for their help.

I am deeply grateful to my wife, Kerstin, for her encouragement and love.

The work on this thesis was financially supported by the German Research Foundation (DFG), grant number TR368/6-1, and the Swiss National Science Foundation (SNSF), grant number 200021-119826/1.

Jan Nesemann

Table of Contents

Introduction

In the theory of quantum mechanics the Hamiltonian H is typically self-adjoint, i.e., $H = H^*$. The self-adjointness ensures that the spectrum of the Hamiltonian, representing the energy spectrum of H, is real but it is not a necessary condition. In the literature on so-called PT-symmetric quantum mechanics (see, e.g., [BB98], [BBM99], [BBJ03], [Ben04b] and [Ben07]), it is believed that self-adjointness is rather a mathematical requirement than a physically established fact. Therefore, it was considered a surprise that operators exist which are not self-adjoint in the given quantum mechanical Hilbert space, but have real spectrum and that – if e.g. complex eigenvalues were present – they occurred only in complex conjugate pairs.

From a mathematical point of view, however, this is no surprise at all – provided one is familiar with the theory of self-adjoint operators in spaces with indefinite inner product (Krein spaces). The physical structure found to be the reason for the reality of the spectrum is PT-symmetry (space-time reflection symmetry), which amounts to self-adjointness in some Krein space. A Hamiltonian H is PT-symmetric if it commutes with PT, that is $PTH = HPT$, compare, e.g., [Ben07] and [AT10]. Here P denotes the space reflection (parity) operator and T the time reflection operator. P and T satisfy the relations $P^2 = T^2 = (PT)^2 = I$ and $PT = TP$. If $p = \mathrm{id}/\mathrm{d}x$ and x are the momentum and position operators, then P has the effect

$$p \mapsto -p, \quad x \mapsto -x$$

and T has the effect

$$p \mapsto -p, \quad x \mapsto x, \quad \mathrm{i} \mapsto -\mathrm{i},$$

compare, e.g., [BB98], [BBM99], [BBJ03], [Ben04b] and [Ben07].

In contrast to self-adjointness in Hilbert spaces, PT-symmetry does not necessarily lead to a completely real spectrum. For example, the Hamiltonian

$$H := p^2 + \mathrm{i}x^3$$

is not symmetric in the Hilbert space $L^2(\mathbb{R})$ since the potential is not real-valued. However, the Hamiltonian H is PT-symmetric in the Hilbert space $L^2(\mathbb{R})$:

$$
\begin{aligned}
\left(PTH(PT)^{-1}\right)f(x) &= \left(PTHTP\right)f(x) \\
&= \left(PTHT\right)f(-x) \\
&= \left(PTH\right)\overline{f(-x)} \\
&= \left(PT\right)\left(p^2\overline{f(-x)} + ix^3\overline{f(-x)}\right) \\
&= P\left(p^2 f(-x) - ix^3 f(-x)\right) \\
&= p^2 f(x) - i(-x)^3 f(x) = Hf(x), \quad f \in \mathscr{D}(H),
\end{aligned}
$$

where $\mathscr{D}(H)$ is the maximal domain of H. More generally, for the family of PT-symmetric Hamiltonians (compare, e.g., [BB98] and [AT10])

$$
H_\varepsilon := p^2 + x^2(ix)^\varepsilon, \quad \varepsilon \in \mathbb{R},
$$

the spectrum of H_ε was found to be real and positive if $\varepsilon \geq 0$ and partly real and partly complex if $\varepsilon < 0$ (see, e.g., [DDT01a] for a proof of the reality of the spectrum for $\varepsilon \geq 0$; for $\varepsilon < 0$ numerical results indicate the appearance of complex eigenvalues, see, e.g., [BB98] and [Ben04a]). More precisely, for $-1 < \varepsilon < 0$, there is a finite number of real positive eigenvalues and an infinite number of complex conjugate pairs of eigenvalues, if $\varepsilon \leq -1$, then there are no real eigenvalues, see, e.g., [BB98] and [Ben04a].

During the last decade PT-symmetric models have been analysed intensively, see, e.g., the review paper [Ben07] and the references therein. Within the vast literature on PT-symmetric problems there are only some mathematically rigorous papers, see, e.g. [DDT01b], [Shi02], [AK04], [LT04], [Shi04], [Shi05], [Tan06], [Tan07] and [AT10]. In particular, we mention the works of E. Caliceti, F. Cannata, S. Graffi and J. Sjöstrand (see [Cal04], [CGS05], [CG05], [Cal05], [CCG06], [CG08] and [CCG08]), who use perturbation theory for linear operators. In [Mos02], [Jap02], [AK04], [LT04], [GSZ05] and [Tan06] Krein space methods were applied to PT-symmetric problems. The paper by H. Langer and C. Tretter (see [LT04] and [LT06]) was the first where Krein space methods were used to prove rigorous abstract results for PT-symmetric problems; this approach is also crucial for this thesis.

Consider the following situation. If a self-adjoint operator A_0 in a Krein space $\left(\mathscr{K}, [\cdot,\cdot]\right)$, which is also self-adjoint with respect to some Hilbert space inner product $(\cdot,\cdot)$ on $\mathscr{K}$, has an isolated real eigenvalue of definite type, then this eigenvalue remains real under a "sufficiently small" PT-symmetric perturbation V (that is, V is symmetric in $(\mathcal{K}, [\cdot, \cdot])$). This theorem has been proven for the case of bounded V in [LT04] and relies on the fact that a uniformly positive subspace of a Krein space is stable, which is a well-known result in the theory of Krein spaces. The theorem mentioned above can be applied to isolated eigenvalues of PT-symmetric problems. If two simple real eigenvalues of the same type meet, they remain real after crossing. This is the case of self-adjoint operators in Hilbert spaces, where all eigenvalues are of positive type. If two real eigenvalues of different type meet, they will, in general, develop into a pair of non-real complex conjugate eigenvalues.

While the case of bounded V was treated in [LT04] (see also [LT06]), a comparable result for the case of unbounded V has been missing. The case of unbounded potentials has only been considered for a few special classes or examples of operators, see, e.g., [DDT01b], [Shi02], [CG05], [Cal05] and [CG08]. The aim of this thesis is to generalize the results of [LT04] to wide classes of unbounded potentials, e.g., to relatively bounded and relatively form-bounded operators. This includes a generalization of the results obtained in [CG05] for a special class of Schrödinger operators with relatively bounded complex polynomial potentials.

The main results of this thesis are stability results for the reality of the spectrum of a family of operators A_ε of the form

$$A_\varepsilon := A_0 + \varepsilon V, \quad \varepsilon \in [0, 1];$$

in particular, we consider the case where A_ε is self-adjoint in a Krein space $(\mathcal{K}, [\cdot, \cdot])$ while A_0 is also self-adjoint with respect to some Hilbert space inner product $(\cdot, \cdot)$ on $\mathcal{K}$. Furthermore, we give inclusions for the perturbed spectrum of A_ε. We found different assumptions on V to prove the respective results. More precisely, we consider the following three types of assumptions on V; in any case V is assumed to be symmetric in the Krein space $(\mathcal{K}, [\cdot, \cdot])$.

(a) $\mathscr{D}(A_0) \subset \mathscr{D}(V)$ and there exist constants $\alpha \geq 0$, $0 \leq \beta < 1/2$ such that

$$(1) \qquad \|Vx\| \leq \alpha\|x\| + \beta\|A_0 x\|, \quad x \in \mathscr{D}(A_0);$$

in this case $A_\varepsilon = A_0 + V$ is defined as an operator sum.

(b) A_0 and V are bounded from below in $(\mathcal{K}, [\cdot, \cdot])$, $\mathscr{D}(\mathfrak{a}_0) \subset \mathscr{D}(\mathfrak{v})$ for the quadratic forms $\mathfrak{a}_0$ and $\mathfrak{v}$ associated with A_0 and V, respectively, and there exist constants $\alpha \geq 0$, $0 \leq \beta < 1/2$ such that

$$|\mathfrak{v}[x]| \leq \alpha\|x\|^2 + \beta|\mathfrak{a}_0[x]|, \quad x \in \mathscr{D}(\mathfrak{a}_0);$$

in this case $A_\varepsilon = A_0 \dotplus V$ is defined as a form sum, which is an extension of the operator sum.

(c) $\mathcal{D}(V) \subset \mathcal{D}(A_0)$ and there exist constants $\alpha \geq 0$, $0 \leq \beta < 1/2$ such that

$$|[Vx,x]| \leq \alpha\|x\|^2 + \beta\,[J|A_0|x,x], \quad x \in \mathcal{D}(V),$$

where J denotes a fundamental symmetry on $\mathcal{K}$, and $\mathcal{D}(V)$ is a core of $|A_0|^{1/2}$; in this case A_ε is the pseudo-Friedrichs extension of $A_0 + \varepsilon V$.

For example, in terms of relative boundedness properties of V with respect to A_0, case (a), our main results are the following. Since, by assumption, A_0 is self-adjoint in a Hilbert space, its spectrum is real. We establish the following conditions which guarantee the spectrum of $A_0 + V$ to be real, even when $A_0 + V$ is not self-adjoint in a Hilbert space (compare Theorem 1.44 below):

(i) Suppose λ^0 is an isolated eigenvalue of A_0 of definite type with finite multiplicity m. If

(2) $$\frac{1}{\delta}\bigl(\alpha + \beta(\delta + |\lambda^0|)\bigr) < \frac{1}{2},$$

where $\delta := \operatorname{dist}\bigl(\lambda^0, \sigma(A_0)\backslash\{\lambda^0\}\bigr)$, then $\sigma(A_1) \cap B_{\delta/2}(\lambda^0)$ consists of a finite system of isolated and real eigenvalues with total multiplicity m which are of the same type as λ^0.

(ii) The preceding result can be extended to the case when the spectrum of A_0 is discrete and consists of an infinite sequence of eigenvalues

$$\cdots < \lambda^0_{-2} < \lambda^0_{-1} < \lambda^0_1 < \lambda^0_2 < \cdots$$

of definite type with finite multiplicities. In this case it is necessary that (2) holds for each eigenvalue λ^0_n, $n \in \mathbb{Z}^*$. Let $\delta_n := \operatorname{dist}\bigl(\lambda^0_n, \sigma(A_0)\backslash\{\lambda^0_n\}\bigr)$, $n \in \mathbb{Z}^*$, and suppose that (1) holds with $\alpha_n \geq 0$ and $\beta_n \in [0, 1/2)$, $n \in \mathbb{Z}^*$, such that

(3) $$\gamma := \sup_{n \in \mathbb{Z}^*}\left(\frac{1}{\delta_n}\bigl(\alpha_n + \beta_n(\delta_n + |\lambda^0_n|)\bigr)\right) < \infty.$$

Then the spectrum of A_ε is discrete and consists of real eigenvalues which are of definite type for all $\varepsilon \in [0, \varepsilon_0]$, where $\varepsilon_0 \in (0, 1]$ has to be chosen such that $\varepsilon_0 < 1/(2\gamma)$.

The preceding result can be illustrated by the following example (see Section 3.3 below) which was first studied in [CG05]. Consider operators induced by the differential expression

$$A_\varepsilon = -\frac{d^2}{dx^2} + P + \varepsilon i Q, \quad \varepsilon \in [0,1],$$

in $L^2(\mathbb{R})$, where P and Q are multiplication operators by real polynomials P and Q; P is an even polynomial of degree $2p$, $p \geq 1$, with $\lim_{|x| \to \infty} P(x) = \infty$, and Q is an odd polynomial of degree $2q - 1$, $q \geq 1$, such that $p > 2q$. In this special case the assumptions of (ii) are satisfied for

$$A_0 = -\frac{d^2}{dx^2} + P \quad \text{and} \quad V = iQ;$$

the spectrum of A_0 consists of an infinite sequence of eigenvalues $\lambda_1^0 < \lambda_2^0 < \cdots$ and the constants $\alpha_n \geq 0$ and $\beta_n \in [0, 1/2)$, $n \in \mathbb{N}$, in (1) can be chosen such that (3) holds.

The results (i) and (ii) can be extended to the case where isolated compact parts of the spectrum of A_0 are considered instead of isolated eigenvalues (see Theorem 1.46 below). Furthermore, the results remain valid for cases (b) of relatively form-bounded operators and (c) of pseudo-Friedrichs extensions.

The proof of the results (i) and (ii) relies on the fact that a uniformly positive subspace of a Krein space is stable (see [LT04, Theorem 3.1]). This stability theorem applies to isolated eigenvalues or isolated (compact) parts of the spectrum of the operator family A_ε. In order to ensure that isolated eigenvalues or isolated parts of the spectrum of A_0 remain isolated under the perturbation εV, it is necessary that the perturbation εV is "sufficiently small" or, equivalently, A_ε is "sufficiently close" to A_0. While the "distance" between two bounded linear operators can be defined as the norm of their difference, the "distance" between two unbounded linear operators has to be measured in a different way. To this end the notion of generalized convergence is used, which amounts to convergence between the graphs of two unbounded linear operators or, equivalently, to the convergence of the resolvent of A_ε to the resolvent of A_0 in norm. The latter is guaranteed by assuming that V is relatively bounded (or relatively form-bounded, respectively) with respect to A_0 with relative bound (relative form-bound, respectively) less than 1.

The results achieved in this thesis are new in various aspects. In cases (a) and (b), results were known only for very particular classes of differential operators (compare [CG05] and [CG08], respectively). For case (c) the results of this thesis have been shown before in [Ves72a] and [Ves72b], but the proofs were different. In comparison to our results, [Ves72b] requires further assumptions but shows in addition to the reality of the spectrum of the pseudo-Friedrichs extension A_1, A_1 is similar to a self-adjoint operator in a Hilbert space, compare Remark 2.52 below.

The thesis is organized as follows. In Chapter 1 the case (a) of relatively bounded V is considered. The first section of this chapter gives a brief introduction into the theory of linear operators in Krein spaces. Subsequently, the reader is provided with fundamental definitions as well as elementary facts for relatively bounded operators. In Section 3 we introduce the notion of generalized convergence and we present a proof of the well-known result that, for an arbitrary family of closed linear operators T_ε, $\varepsilon \in [0,1]$, in a Banach space, T_ε converges to T_0 in the generalized sense if and only if the resolvent of T_ε converges to the resolvent of T_0 in norm. Further, we recall important results from perturbation theory regarding the change of the spectrum. If a Cauchy contour Γ separates a bounded part of the spectrum $\sigma(T_0)$ of T_0 from the rest and T_ε converges to T_0 in the generalized sense, the spectrum of T_ε is likewise separated into two parts by Γ, moreover, the isolated part enclosed by Γ changes continuously with ε. If V is A_0-bounded with A_0-bound less than 1, then A_ε converges to A_0 in the generalized sense and hence the above results apply to the family of operators $A_\varepsilon = A_0 + \varepsilon V$. Consequently, isolated eigenvalues or isolated parts of the spectrum of A_0 remain isolated under the perturbation εV. This enables us to apply the Krein space methods of [LT04] to establish criteria for the operator A_ε to have real spectrum consisting of isolated eigenvalues or isolated parts if this holds for A_0.

Chapter 2 extends the results of Chapter 1 to the case (b) of relatively form-bounded perturbations. Instead of studying the usual operator sum $A_0 + \varepsilon V$, we consider the sum $A_0 \dotplus \varepsilon V$ of A_0 and εV defined by means of quadratic forms which is an extension of the operator sum $A_0 + \varepsilon V$. While the condition of relative form-boundedness itself is less restrictive than the one of relative boundedness, relatively form-bounded operators have to be required to be bounded from below (with respect to the respective inner product); therefore, case (b) constitutes a different class of unbounded perturbations compared to case (a). Nevertheless, as in case (a), relative form-boundedness of V with respect to A_0 with relative form-bound less than 1 guarantees that A_ε converges to A_0 in the generalized sense. This enables us to extend the results of Chapter 1 to the case of relatively form-bounded operators.

At the end of the second chapter, we consider case (c) and introduce the notion of pseudo-Friedrichs extensions. A pseudo-Friedrichs extension is an extension of the usual operator sum which is different from the form-sum introduced before; in particular, the domain inclusion is $\mathcal{D}(V) \subset \mathcal{D}(A_0)$ rather than $\mathcal{D}(A_0) \subset \mathcal{D}(V)$ (case (a)) or $\mathcal{D}(\mathfrak{a}_0) \subset \mathcal{D}(\mathfrak{v})$ (case (b)). The results are not essentially related to sesquilinear forms, but the techniques used in the proofs are similar. In the context of Krein spaces, these operators have also been studied in [Ves72a], [Ves72b] and [Ves08] where similar results were obtained, but by different proofs.

In Chapter 3 we present some examples where the results of this thesis are applied to ordinary differential operators. In Sections 3.1 and 3.2 we study a second and a fourth order differential operator, respectively, on a compact interval. The class of differential operators on $\mathbb{R}$ introduced in [CG05] which is also covered by the results of this thesis is considered in Section 3.3. For all these examples the results show that the spectrum of $A_0 + V$ remains real, even though $A_0 + V$ is not self-adjoint in a Hilbert space.

Notation. For an introduction to the theory of unbounded linear operators, the following notation and basic terminology as well as for further details we refer to [Kat95], [GGK90] and [RS80, RS75, RS79, RS78]. The domain of a linear operator A in a Banach space $\mathcal{X}$ we denote by $\mathcal{D}(A)$, the range of A by $\mathcal{R}(A)$ and the graph of A by $\mathcal{G}(A)$. If A is a closed linear operator, we denote the spectrum and the resolvent set of A by $\sigma(A)$ and $\rho(A)$, respectively.

Chapter 1

Relatively Bounded Perturbations in Krein Spaces

1.1 Linear Operators in Krein Spaces

The main results of this thesis are based on the theory of linear operators acting in Krein spaces. In this section we briefly outline the definitions and some elementary facts. A detailed study of Krein spaces and linear operators therein can be found in [Lan62], [Lan82], [AI89], [Bog74] and [And79].

Definition 1.1. An inner product space $(\mathcal{K}, [\cdot,\cdot])$ is called **Krein space** if it contains two subspaces $\mathcal{H}_+$, $\mathcal{H}_-$ with the following properties:

$$(1.1) \qquad \mathcal{K} = \mathcal{H}_+[\dotplus]\mathcal{H}_-,$$

$$(1.2) \qquad (\mathcal{H}_+, [\cdot,\cdot]) \text{ and } (\mathcal{H}_-, -[\cdot,\cdot]) \text{ are Hilbert spaces;}$$

here $[\dotplus]$ denotes the direct $[\cdot,\cdot]$-orthogonal sum.

Condition (1.2) means that the inner product is positive definite on $\mathcal{H}_+$ and negative definite on $\mathcal{H}_-$, i.e., $[x,x] > 0$ for $x \in \mathcal{H}_+$, $x \neq 0$, $[x,x] < 0$ for $x \in \mathcal{H}_-$, $x \neq 0$, and that $\mathcal{H}_+$ and $\mathcal{H}_-$ are complete with respect to the norms $\|x\| = [x,x]^{1/2}$, $x \in \mathcal{H}_+$ and $\|x\| = (-[x,x])^{1/2}$, $x \in \mathcal{H}_-$, respectively.

According to this definition, the class of Krein spaces includes Hilbert spaces for which $\mathcal{H}_- = \{0\}$. The decomposition (1.1) is not unique and each decomposition (1.1) defines a Hilbert space inner product on $\mathcal{K}$ by the relation

$$(1.3) \qquad (x,y) := [x_+,y_+] - [x_-,y_-], \quad x = x_+ + x_-, y = y_+ + y_-, \quad x_\pm, y_\pm \in \mathcal{H}_\pm.$$

Although these inner products depend on the decomposition (1.1) the norms generated by them are all equivalent (see [Lan82, Proposition I.1.2]). Any of these norms is denoted by $\|\cdot\|$, and so is the corresponding operator norm. If no other topology is explicitly mentioned, all the topological notions refer to this Hilbert space norm topology. For example, a subspace of $\mathscr{K}$ is a linear manifold in $\mathscr{K}$ which is closed with respect to this Hilbert space topology, continuity of operators means continuity with respect to this Hilbert space topology et cetera.

Definition 1.2. Let $(\mathscr{K},[\cdot,\cdot])$ be a Krein space. A bounded linear operator J in $\mathscr{K}$ such that $Jx := x_+ - x_-$ if $x = x_+ + x_-, x_\pm \in \mathscr{H}_\pm$, is called **fundamental symmetry** corresponding to a decomposition (1.1).

The relationship between a Krein space $(\mathscr{K},[\cdot,\cdot])$ and the Hilbert space $(\mathscr{K},(\cdot,\cdot))$ arising from $\mathscr{K}$ using any of the equivalent Hilbert space inner products $(\cdot,\cdot)$ defined by (1.3) can be described as follows. We introduce the linear operators

$$P_\pm x := x_\pm \quad \text{if} \quad x = x_+ + x_-, \, x_\pm \in \mathscr{H}_\pm,$$

which are the $(\cdot,\cdot)$-orthogonal projections onto $\mathscr{H}_+$ and $\mathscr{H}_-$, respectively. Then we have, for the fundamental symmetry $J = P_+ - P_-$ corresponding to (1.1),

(1.4) $[x,y] = [x_+,y_+] - [-x_-,y_-] = (x_+ - x_-, y_+ + y_-) = (Jx,y)$

as well as

(1.5) $(x,y) = [x_+,y_+] + [-x_-,y_-] = [x_+ - x_-, y_+ + y_-] = [Jx,y]$

for $x = x_+ + x_-$ and $y = y_+ + y_-$ with $x_\pm, y_\pm \in \mathscr{H}_\pm$. Furthermore, $J^2 = P_+ + P_- = I$ and $J = J^* = J^{-1}$, where * denotes the adjoint with respect to the Hilbert space inner product (1.3).

Remark 1.3. In the following, for a given Krein space $(\mathscr{K},[\cdot,\cdot])$ by $(\mathscr{K},(\cdot,\cdot))$ we always denote the Hilbert space corresponding to $(\mathscr{K},[\cdot,\cdot])$ as described above.

Definition 1.4. Let $(\mathscr{K},[\cdot,\cdot])$ be a Krein space. An element $x \in \mathscr{K}$ is called
  (i) **positive** if $[x,x] > 0$,
  (ii) **negative** if $[x,x] < 0$ and
  (iii) **neutral** if $[x,x] = 0$.

A subspace $\mathscr{L}$ of $\mathscr{K}$ is called **non-negative** (**non-positive**, respectively) if all elements of $\mathscr{L}$ are not negative (not positive, respectively); $\mathscr{L}$ is said to be **positive** (**negative**, respectively) if all the non-zero elements of $\mathscr{L}$ are positive (negative, respectively), and $\mathscr{L}$ is called **neutral** if $[x,x] = 0$ for all $x \in \mathscr{L}$. Further, $\mathscr{L}$ is called **uniformly positive** if there exists a constant $\gamma > 0$ such that

$$[x,x] \geq \gamma \|x\|^2, \quad x \in \mathscr{L},$$

and, analogously, **uniformly negative** if there exists a constant $\gamma > 0$ such that

$$[x,x] \leq -\gamma \|x\|^2, \quad x \in \mathscr{L}.$$

Here $\| \cdot \|$ denotes any of the equivalent Hilbert space norms generated by a decomposition (1.1). A subspace $\mathscr{L}$ of $\mathscr{K}$ is called **definite** if it is either positive or negative, and **indefinite** if it is neither positive nor negative. The term **uniformly definite** is defined accordingly.

In order to define adjoint operators with respect to the indefinite inner product $[\cdot,\cdot]$, we need the following analogue of F. Riesz' representation theorem, which follows from the latter using (1.4) and (1.5).

Theorem 1.5 (F. Riesz' representation theorem). *Let $(\mathscr{K},[\cdot,\cdot])$ be a Krein space and f a bounded linear functional on $\mathscr{K}$. Then there exists a uniquely determined element $y_f \in \mathscr{K}$ such that*

$$f(x) = [x,y_f], \quad x \in \mathscr{K},$$

and $\|f\| = \|y_f\|$.

Conversely, every $y \in \mathscr{K}$ defines a bounded linear functional f_y on $\mathscr{K}$ by

$$f_y(x) := [x,y], \quad x \in \mathscr{K},$$

with $\|f_y\| = \|y\|$.

Proof. The proof follows from the Hilbert space version of F. Riesz' representation theorem. If f is a bounded linear functional on $\mathscr{K}$, then, since boundedness refers to the Hilbert space norm on $\mathscr{K}$, there exists a unique $y'_f \in \mathscr{K}$ such that $f(x) = (x,y'_f)$, $x \in \mathscr{K}$, and $\|f\| = \|y'_f\|$. Now the claim follows from (1.4) and (1.5) with $y_f := Jy'_f$.

Conversely, let $y \in \mathscr{K}$ be arbitrary and set $y' := Jy \in \mathscr{K}$. Then, by the Hilbert space version of the theorem, $f_y(x) := [x,y] = (x,Jy) = (x,y')$ is a bounded linear functional on $\mathscr{K}$ with $\|f_y\| = \|y'\| = \|y\|$. ∎

Corollary 1.6. *A Krein space* $\left(\mathcal{K},[\cdot,\cdot]\right)$ *is reflexive.*

Proof. Let $y'' \in \mathcal{K}''$, the bidual space of $\mathcal{K}$. According to Theorem 1.5 applied to $\mathcal{K}'$ there exists a $y' \in \mathcal{K}'$ such that for all $x' \in \mathcal{K}'$

$$y''(x') = [x',y']_{\mathcal{K}'} = [y,x]_{\mathcal{K}} = f_x(y) = x'(y) = \left(i_{\mathcal{K}}(y)\right)(x'),$$

where $i_{\mathcal{K}} : \mathcal{K} \to \mathcal{K}''$ is the canonical embedding of $\mathcal{K}$ into its bidual space. That is, $i_{\mathcal{K}} : \mathcal{K} \to \mathcal{K}''$ is surjective. ∎

Definition 1.7. Let $\left(\mathcal{K},[\cdot,\cdot]\right)$ be a Krein space and A a densely defined linear operator in $\mathcal{K}$. Then, the ***Krein space adjoint*** $A^{[*]}$ of A is defined by

$$\mathcal{D}\left(A^{[*]}\right) := \left\{y \in \mathcal{K} : [A\cdot,y] \text{ is continuous on } \mathcal{D}(A)\right\},$$
$$[Ax,y] = [x,A^{[*]}y], \quad x \in \mathcal{D}(A), \; y \in \mathcal{D}\left(A^{[*]}\right).$$

Further, A is called **symmetric** (**in the Krein space** $\mathcal{K}$) if $A \subset A^{[*]}$ and **self-adjoint** if $A = A^{[*]}$.

In the following we always denote the Krein space adjoint (compare, e.g., [Lan82]) by $^{[*]}$ and the Hilbert space adjoint by *. If J denotes a fundamental symmetry corresponding to (1.1), then the Krein space adjoint $A^{[*]}$ is sometimes also called J-***adjoint*** because of the identity

(1.6) $$A^{[*]} = JA^*J$$

(see [Lan62, Lemma I.5] or [Lan82, Paragraph I.3]). Correspondingly, A is also called J-***symmetric*** if $A \subset A^{[*]}$ and J-***self-adjoint*** if $A = A^{[*]}$.

The following lemma gives a convenient relation between self-adjointness of linear operators in Krein and Hilbert spaces.

Lemma 1.8. *Let A be a densely defined linear operator in a Krein space* $\left(\mathcal{K},[\cdot,\cdot]\right)$. *Further let the bounded linear operator J in* $\left(\mathcal{K},[\cdot,\cdot]\right)$ *be a fundamental symmetry. Then A is self-adjoint (symmetric, respectively) with respect to the Krein space inner product $[\cdot,\cdot]$ if and only if JA is self-adjoint (symmetric, respectively) with respect to the Hilbert space inner product $(\cdot,\cdot)$.*

Proof. Since J is bounded, we have $\mathcal{D}\left(JA\right) = \mathcal{D}\left(A\right)$ and $(JA)^* = A^*J^*$ (see, e.g., [MV97, Lemma 19.9]). If $A \subset A^{[*]}$, then, by (1.4),

$$(JAx,y) = [Ax,y] = [x,Ay] = (Jx,Ay) = (x,JAy), \quad x,y \in \mathcal{D}\left(JA\right) = \mathcal{D}\left(A\right),$$

which implies that $JA \subset (JA)^*$, i.e., JA is symmetric with respect to the Hilbert space inner product $(\cdot,\cdot)$ on $\mathscr{K}$. If $A = A^{[*]}$, then, in addition,

$$\mathscr{D}(JA) = \mathscr{D}\big(JA^{[*]}\big) = \mathscr{D}\big(JJA^*J\big) = \mathscr{D}\big(A^*J^*\big) = \mathscr{D}\big((JA)^*\big),$$

i.e., $JA = (JA)^*$.

Vice versa, if $JA \subset (JA)^*$, then

$$[Ax,y] = (JAx,y) = (x,JAy) = [Jx,JAy] = [x,Ay], \quad x,y \in \mathscr{D}(A),$$

which implies that $JA \subset (JA)^{[*]}$, i.e., JA is symmetric with respect to the Krein space inner product $[\cdot,\cdot]$ on $\mathscr{K}$. If $JA = (JA)^*$, then, in addition,

$$\mathscr{D}(A) = \mathscr{D}\big(J^2A\big) = \mathscr{D}\big(J(JA)^*\big) = \mathscr{D}\big(JA^*J^*\big) = \mathscr{D}\big(JA^*J\big) = \mathscr{D}\big(A^{[*]}\big),$$

i.e., $A = A^{[*]}$. ∎

While the spectrum of a self-adjoint operator in a Hilbert space is always real, the spectrum of a self-adjoint operator in a Krein space may be complex. However, it is well-known that it is symmetric with respect to the real axis (see [Lan82, Paragraph I.3]). Since this property is crucial in the following we include its proof. Here and in the sequel, for a subset $\mathscr{M} \subset \mathbb{C}$, we denote by

$$\mathscr{M}^* := \big\{z : \overline{z} \in \mathscr{M}\big\}$$

the set complex conjugate to $\mathscr{M}$.

Theorem 1.9. *Let A be a self-adjoint operator in a Krein space $(\mathscr{K},[\cdot,\cdot])$. Then the spectrum $\sigma(A)$ of A is symmetric with respect to the real axis.*

Proof. Using relation (1.6) we obtain for $\lambda \in \rho(A)$

$$\big((A - \lambda I)^{-1}\big)^{[*]} = J\big((A - \lambda I)^{-1}\big)^* J = J\big((A - \lambda I)^*\big)^{-1} J = \big(J(A - \lambda I)^* J\big)^{-1}$$
$$= \big((A - \lambda I)^{[*]}\big)^{-1} = \big(A^{[*]} - \overline{\lambda} I\big)^{-1},$$

which yields $\sigma(A^{[*]}) = \sigma(A)^*$. Hence $\sigma(A) = \sigma(A)^*$, since A is self-adjoint in $(\mathscr{K},[\cdot,\cdot])$. ∎

Definition 1.10. Let $(\mathscr{K},[\cdot,\cdot])$ be a Krein space and A be a self-adjoint operator in $\mathscr{K}$. A real isolated eigenvalue λ is called of ***positive type*** (***negative type***, respectively) if the corresponding algebraic eigenspace $\mathscr{L}_\lambda$ is positive (negative, respectively), and λ is called **critical** if $\mathscr{L}_\lambda$ is neutral. If an eigenvalue is either of positive or of negative type it is called of **definite type**.

If for an eigenvalue λ the dimension of the corresponding algebraic eigenspace $\mathscr{L}_\lambda$ is finite, then the **multiplicity** of λ is defined as the dimension of $\mathscr{L}_\lambda$. The **total multiplicity** of a finite set of eigenvalues is the sum of all its multiplicities or the dimension of the linear span of all the corresponding algebraic eigenspaces.

1.2 Stability Theorems

In this paragraph the notion of relatively bounded and relatively compact operators in Banach spaces is recalled. A number of important properties of linear operators are preserved under relatively bounded or relatively compact perturbations. For a detailed study of relatively bounded and relatively compact operators see [Kat95], [EE87], [GGK90] and [Gol66].

1.2.1 Relatively Bounded and Relatively Compact Operators

Definition 1.11. Let $\mathscr{X}$, $\mathscr{Y}_1$ and $\mathscr{Y}_2$ be Banach spaces. Further, let A and V be linear operators, A from $\mathscr{X}$ to $\mathscr{Y}_1$ and V from $\mathscr{X}$ to $\mathscr{Y}_2$. If $\mathscr{D}(A) \subset \mathscr{D}(V)$ and there exist non-negative constants α and β such that

$$(1.7) \qquad \|Vx\| \le \alpha\|x\| + \beta\|Ax\|, \quad x \in \mathscr{D}(A),$$

then V is called **relatively bounded with respect to A** or simply **A-bounded**. The greatest lower bound β_0 of all possible constants β in (1.7) is called **relative bound of V with respect to A** or simply **A-bound**, i.e.,

$$\beta_0 := \inf\{\beta \ge 0 : \exists \alpha \ge 0 \text{ such that } \|Vx\| \le \alpha\|x\| + \beta\|Ax\|, x \in \mathscr{D}(A)\}.$$

Clearly, a bounded everywhere defined operator V is relatively bounded with respect to any operator A with relative bound 0.

The converse is not true: in general, it is not possible to choose $\beta = \beta_0$ in inequality (1.7) (compare also [Kat95, Example IV.1.6]).

Remark 1.12. If V is relatively bounded with respect to A with A-bound $\beta_0 \geq 0$, then for any $\beta > \beta_0$ there exists an $\alpha_\beta > 0$ such that

$$(1.8) \qquad \|Vx\| \leq \alpha_\beta \|x\| + \beta \|Ax\|, \quad x \in \mathscr{D}(A).$$

An alternative to inequality (1.7) and to determine the relative bound is given by the following lemma (see [Kat95, Section V.4.1]).

Lemma 1.13. *The following statements are equivalent:*

(i) $\exists\, \alpha, \beta \geq 0$ *such that* $\|Vx\| \leq \alpha\|x\| + \beta\|Ax\|$, $x \in \mathscr{D}(A)$,

(ii) $\exists\, \alpha', \beta' \geq 0$ *such that* $\|Vx\|^2 \leq \alpha'^2\|x\|^2 + \beta'^2\|Ax\|^2$, $x \in \mathscr{D}(A)$.

Moreover,

$$(1.9) \quad \begin{aligned} \beta_0 &= \inf\{\beta \geq 0 : \exists\, \alpha \geq 0 \text{ such that } \|Vx\| \leq \alpha\|x\| + \beta\|Ax\|, x \in \mathscr{D}(A)\} \\ &= \inf\{\beta' \geq 0 : \exists\, \alpha' \geq 0 \text{ such that } \|Vx\|^2 \leq \alpha'^2\|x\|^2 + \beta'^2\|Ax\|^2, x \in \mathscr{D}(A)\}. \end{aligned}$$

Proof. If (ii) holds, then, obviously,

$$\|Vx\|^2 \leq \alpha'^2\|x\|^2 + \beta'^2\|Ax\|^2 \leq \left(\alpha'\|x\| + \beta'\|Ax\|\right)^2, \quad x \in \mathscr{D}(A),$$

and so (i) holds with $\alpha := \alpha'$, $\beta := \beta'$.

Vice versa, suppose that (i) holds and let $\varepsilon > 0$ be arbitrary. Then we have the estimates

$$\|Vx\|^2 \leq \alpha^2\|x\|^2 + \beta^2\|Ax\|^2 + 2\alpha\beta\|x\|\|Ax\|$$

$$\leq \alpha^2\|x\|^2 + \beta^2\|Ax\|^2 + 2\alpha\beta\|x\|\|Ax\| + \left(\frac{\alpha}{\sqrt{\varepsilon}}\|x\| - \sqrt{\varepsilon}\beta\|Ax\|\right)^2$$

$$= \alpha^2\|x\|^2 + \beta^2\|Ax\|^2 + \frac{\alpha^2}{\varepsilon}\|x\|^2 + \varepsilon\beta^2\|Ax\|^2$$

$$= \left(1 + \frac{1}{\varepsilon}\right)\alpha^2\|x\|^2 + (1+\varepsilon)\beta^2\|Ax\|^2, \quad x \in \mathscr{D}(A).$$

Hence (ii) holds with $\alpha' := \sqrt{\left(1 + \frac{1}{\varepsilon}\right)\alpha^2} \geq 0$ and $\beta' := \sqrt{(1+\varepsilon)\beta^2} \geq 0$. Since $\varepsilon > 0$ is arbitrary, the A-bound of V may as well be defined as the greatest lower bound of the possible values of β'. ∎

Definition and Remark 1.14. Let $\mathscr{X}$ and $\mathscr{Y}$ be Banach spaces and let A be a closed linear operator from $\mathscr{X}$ to $\mathscr{Y}$. Set

$$\|x\|_A := \|x\| + \|Ax\|, \quad x \in \mathscr{D}(A).$$

Then $\|\cdot\|_A$ defines a norm on $\mathscr{D}(A)$ which is called **graph norm** (and some-times also **A-norm**). Further, $\big(\mathscr{D}(A), \|\cdot\|_A\big)$ becomes a Banach space, which we denote by $\mathscr{D}_A$.

Proof. Indeed $\|\cdot\|_A$ defines a norm on $\mathscr{D}(A)$ since $\|\cdot\|$ is a norm and A is linear. By assumption, A is closed, i.e., given a sequence $(x_n)_{n=1}^{\infty} \in \mathscr{D}(A)$ such that $x_n \to x$ and $Ax_n \to y$, it follows that $x \in \mathscr{D}(A)$ and $Ax = y$. Hence each Cauchy sequence in $\mathscr{D}(A)$ converges in $\mathscr{D}(A)$, i.e., $\big(\mathscr{D}(A), \|\cdot\|_A\big)$ is a Banach space. ∎

Remark 1.15. Let $\mathscr{X}$, $\mathscr{Y}_1$ and $\mathscr{Y}_2$ be Banach spaces and let A be a closed linear operator from $\mathscr{X}$ to $\mathscr{Y}_1$. Suppose that V is a linear operator from $\mathscr{X}$ to $\mathscr{Y}_2$ such that $\mathscr{D}(A) \subset \mathscr{D}(V)$. Then the restriction of V to $\mathscr{D}(A)$ can be regarded as a linear operator $\widehat{V}$ from $\mathscr{D}_A$ to $\mathscr{Y}_2$. By Definition and Remark 1.14, $\widehat{V}$ is bounded if and only if V is A-bounded.

Proof. Let V be A-bounded and let α, β such that (1.7) holds. Then

$$(1.10) \qquad \big\|\widehat{V}x\big\| = \|Vx\| \le \alpha\|x\| + \beta\|Ax\| \le \max\{\alpha,\beta\}\|x\|_A, \quad x \in \mathscr{D}(A).$$

Vice versa, if $\widehat{V}$ is bounded, then

$$\|Vx\| = \big\|\widehat{V}x\big\| \le \big\|\widehat{V}\big\|\|x\|_A = \big\|\widehat{V}\big\|\|x\| + \big\|\widehat{V}\big\|\|Ax\|, \quad x \in \mathscr{D}(A).$$

Hence (1.7) holds with $\alpha := \beta := \big\|\widehat{V}\big\|$. ∎

Remark 1.16. If A is closed and V is closable, then $\mathscr{D}(A) \subset \mathscr{D}(V)$ implies that V is A-bounded.

Proof. Let $\widehat{V}$ be defined as in Remark 1.15. According to Remark 1.15, V is A-bounded if and only if $\widehat{V}$ is bounded in $\mathscr{D}_A$. It is thus sufficient to show that $\widehat{V}$ is closed in $\mathscr{D}_A$, since then, by the Closed Graph Theorem, $\widehat{V}$ is bounded in $\mathscr{D}_A$. Let $(x_n)_{n=0}^{\infty} \subset \mathscr{D}(A)$ such that $x_n \overset{\|\cdot\|_A}{\to} 0$ and $\widehat{V}x_n \to y$ for some $y \in \mathscr{Y}_2$. Since $x_n \overset{\|\cdot\|_A}{\to} 0$ also $x_n \overset{\|\cdot\|}{\to} 0$. Hence, since V is closable, $\widehat{V}x_n = Vx_n \to 0$, that is, $\widehat{V}$ is closable. Consequently, $\widehat{V}$ is closed since $\widehat{V}$ is defined on all of $\mathscr{D}(A) = \mathscr{D}_A$. ∎

We note the following obvious relation.

Remark 1.17. Let A and V be linear operators in a Krein space $\mathscr{K}$. Then V is A-bounded with A-bound β_0 if and only if JV is JA-bounded with JA-bound β_0.

Proof. The proof follows from the identity $\|Jx\| = \|x\|$, $x \in \mathcal{K}$. ∎

Definition 1.18. Let $\mathcal{X}$, $\mathcal{Y}_1$ and $\mathcal{Y}_2$ be Banach spaces. Further, let A and V be linear operators, A from $\mathcal{X}$ to $\mathcal{Y}_1$ and V from $\mathcal{X}$ to $\mathcal{Y}_2$. If $\mathcal{D}(A) \subset \mathcal{D}(V)$ and for any sequence $(x_n)_{n=1}^{\infty} \subset \mathcal{D}(A)$ such that $(x_n)_{n=1}^{\infty}$ and $(Ax_n)_{n=1}^{\infty}$ are bounded, $(Vx_n)_{n=1}^{\infty}$ has a convergent subsequence, then V is called *relatively compact with respect to* A or simply A-*compact*.

Remark 1.19. If V is A-compact, then V is A-bounded.

Proof. Assume V is not A-bounded. Then there exists a sequence $(x_n)_{n=1}^{\infty} \subset \mathcal{D}(A)$ such that $\|x_n\|_A = \|x_n\| + \|Ax_n\| = 1$ but $\|Vx_n\| \geq n$ for all $n \in \mathbb{N}$. Since $(Vx_n)_{n=1}^{\infty}$ does not have a convergent subsequence, which is obvious, this is a contradiction to the A-compactness of V. ∎

Remark 1.20. Let $\mathcal{X}$, $\mathcal{Y}_1$ and $\mathcal{Y}_2$ be Banach spaces and let A be a closed linear operator from $\mathcal{X}$ to $\mathcal{Y}_1$. Suppose that V is a linear operator from $\mathcal{X}$ to $\mathcal{Y}_2$ such that $\mathcal{D}(A) \subset \mathcal{D}(V)$. Then the restriction $\widehat{V}$ of V to $\mathcal{D}(A)$ defined as in Remark 1.15 is compact if and only if V is A-compact.

Proof. The claim is obvious since $(x_n)_0^{\infty}$ is bounded in $\mathcal{D}_A$ if and only if $(x_n)_0^{\infty}$ and $(Ax_n)_0^{\infty}$ are bounded in $\mathcal{X}$ and $\mathcal{Y}_1$, respectively, and $\widehat{V} = V|_{\mathcal{D}(A)}$. ∎

1.2.2 The Case of Relative Bound 0

In the following we are particularly interested in relatively bounded operators having relative bound 0. Necessary conditions for this special case can be found in the literature, see [Hes69], [Wei00, Sec. 9.2], [EE87, Sec. III.7] and [Gol66, Sec. V.3].

A Hilbert space version of the following lemma and its proof can be found in [Jör67], see Hilfssatz 1.1 therein.

Lemma 1.21. *Let $\mathcal{X}$ and $\mathcal{Y}$ be Banach spaces and let A and V be linear operators from $\mathcal{X}$ to $\mathcal{X}$ and $\mathcal{X}$ to $\mathcal{Y}$, respectively, such that $\mathcal{D}(A) \subset \mathcal{D}(V)$ and V is relatively bounded with respect to A. If, for each sequence $(x_n)_{n=1}^{\infty} \subset \mathcal{D}(A)$, the sequences $(x_n)_{n=1}^{\infty}$ and $(Ax_n)_{n=1}^{\infty}$ converge weakly to zero, then also $(Vx_n)_{n=1}^{\infty}$ converges weakly to zero.*

Proof. Let g be a bounded linear functional on $\mathscr{Y}$. Define a linear functional f on the graph $\mathscr{G}(A) := \{(x, Ax) : x \in \mathscr{D}(A)\}$ of A by $f\big((x, Ax)\big) = g(Vx)$ for $x \in \mathscr{D}(A)$. Since V is A-bounded, we have

$$\big|f\big((x, Ax)\big)\big|^2 = |g(Vx)|^2 \leq \|g\|^2 \|Vx\|^2 \leq \|g\|^2 \max\{\alpha', \beta'\}^2 \|(x, Ax)\|^2,$$

that is, f is bounded on $\mathscr{G}(A)$. According to the Hahn-Banach-Theorem, there exists a bounded linear functional $\widetilde{f}$ on $\mathscr{X} \oplus \mathscr{X}$, such that $\widetilde{f} = f$ on $\mathscr{G}(A)$. Consequently,

$$(1.11) \qquad g(Vx) = f\big((x, Ax)\big) = \widetilde{f}\big((x, Ax)\big) = \widetilde{f}\big((x, 0)\big) + \widetilde{f}\big((0, Ax)\big).$$

If for some $(x_n)_{n=1}^{\infty} \subset \mathscr{D}(A)$ the sequences $(x_n)_{n=1}^{\infty}$ and $(Ax_n)_{n=1}^{\infty}$ converge weakly to zero, then, by equation (1.11), also $(Vx_n)_{n=1}^{\infty}$ converges weakly to zero. ∎

The statements and proofs of the following theorem can be found in [Wei00, Satz 9.13] and [Hes69], respectively.

Theorem 1.22. *Let $\mathscr{X}$, $\mathscr{Y}_1$ and $\mathscr{Y}_2$ be Banach spaces and let A and V be linear operators from $\mathscr{X}$ to $\mathscr{Y}_1$ and $\mathscr{X}$ to $\mathscr{Y}_2$, respectively, such that $\mathscr{D}(A) \subset \mathscr{D}(V)$ and V is relatively compact with respect to A. Then V has A-bound 0 if*

(i) *V is closable, or*

(ii) *$\mathscr{Y}_2$ is reflexive and A is closable.*

Proof. By Remark 1.19, V is A-bounded. We prove the theorem by contradiction. Suppose that the A-bound of V is not 0. Then there exist an $\varepsilon > 0$ and a sequence $(x_n)_{n=1}^{\infty} \subset \mathscr{D}(A)$ such that for all natural numbers n

$$(1.12) \qquad\qquad \|Vx_n\| > n\|x_n\| + \varepsilon\|Ax_n\|.$$

Set $y_n := x_n / \|x_n\|_A$. Then, by inequalities (1.7), (1.10) and (1.12), and since $\|y_n\|_A = 1$, $n \in \mathbb{N}$, we have for $n \geq \varepsilon$

$$(1.13) \quad \max\{\alpha, \beta\} = \max\{\alpha, \beta\}\|y_n\|_A \geq \|Vy_n\| > n\|y_n\| + \varepsilon\|Ay_n\| \geq \varepsilon\big(\|y_n\| + \|Ay_n\|\big),$$

where α and β are constants according to (1.7). Hence, by inequality (1.13), $y_n \to 0$ if $n \to \infty$ and $(y_n)_{n=1}^{\infty}$ and $(Ay_n)_{n=1}^{\infty}$ are bounded sequences. Thus, the A-compactness of V implies the existence of a subsequence $(y_{n_k})_{k=1}^{\infty}$ of $(y_n)_{n=1}^{\infty}$ such that $(Vy_{n_k})_{k=1}^{\infty}$ converges to some $z \in \mathscr{Y}_2$. In the following we show that, if either (i) or (ii) holds, then $z = 0$.

Let V be closable. Then, since $(y_{n_k})_{k=1}^{\infty}$ converges to zero and $(Vy_{n_k})_{k=1}^{\infty}$ converges to z, it follows that $z = 0$.

Now let $\mathscr{Y}_2$ be reflexive and A be closable. Since $\mathscr{Y}_2$ is reflexive and $(Ay_{n_k})_{k=1}^{\infty}$ is bounded, there exists, by [Wer07, Theorem III.3.7], a subsequence $(y_{n_{k_l}})_{l=1}^{\infty}$ of $(y_{n_k})_{k=1}^{\infty}$ such that $(Ay_{n_{k_l}})_{l=1}^{\infty}$ converges weakly to some $v \in \mathscr{Y}_2$. We have

$$(0,v) = \text{w} - \lim_{l \to \infty} (y_{n_{k_l}}, Ay_{n_{k_l}}) \in \overline{\mathscr{G}(A)} = \mathscr{G}(\overline{A})$$

and hence $v = 0$. According to Lemma 1.21 it follows that also $(Vy_{n_{k_l}})_{l=1}^{\infty}$ converges weakly to zero. Since we already know that $(Vy_{n_k})_{k=1}^{\infty}$ converges to z, it follows that $z = 0$.

Altogether, since we have shown that $(y_{n_k})_{k=1}^{\infty}$ and $(Vy_{n_k})_{k=1}^{\infty}$ converge to 0, (1.13) implies that so does $(Ay_{n_k})_{k=1}^{\infty}$. This leads to the contradiction

$$1 = \|y_{n_k}\|_A = \|y_{n_k}\| + \|Ay_{n_k}\| \longrightarrow 0, \quad k \to 0. \qquad \blacksquare$$

We conclude the following:

Remark 1.23. Let $(\mathscr{K}, [\cdot, \cdot])$ be a Krein space and let A and V be linear operators in $(\mathscr{K}, [\cdot, \cdot])$ such that V is A-compact. Then V has A-bound 0 if at least one of the operators A or V is closable.

Sufficient conditions for relatively bounded operators with relative bound 0 to be relatively compact can be found in [Hes69] and [Wei00, Section 9.2].

1.2.3 Stability of Self-Adjointness in Krein Spaces

When a self-adjoint operator A in a Krein space is perturbed by a symmetric operator V which is relatively bounded with respect to A, then, in general, the perturbed operator $A + V$ need not to be self-adjoint in the Krein space. In the Hilbert space case, a sufficient condition is that the A-bound of V is less than 1. This so called Kato-Rellich theorem extends to the case of Krein spaces.

Theorem 1.24 (Kato-Rellich for Krein spaces). *Let A be a self-adjoint operator in a Krein space $(\mathscr{K}, [\cdot, \cdot])$. If V is a symmetric and A-bounded operator in $(\mathscr{K}, [\cdot, \cdot])$ with A-bound less than 1, then $A + V$ is self-adjoint in $(\mathscr{K}, [\cdot, \cdot])$.*

Proof. With the help of Lemma 1.8 and Remark 1.17 the proof follows from the Hilbert space version of the theorem, see, e.g., [Kat95, Theorem V.4.3]. ∎

The Kato-Rellich theorem for Krein spaces gives a very important result regarding the structure of the spectrum of the perturbed operator $A + V$. If $A + V$ is self-adjoint in $(\mathcal{K},[\cdot,\cdot])$, its spectrum is symmetric with respect to the real axis. In particular, non-real eigenvalues occur as complex conjugate pairs.

1.3 Continuity of Separated Parts of the Spectrum

In this section we consider relatively bounded perturbations of closed linear operators and their effect on the spectrum. If such an operator A_0 is perturbed, e.g., by a closed linear operator V which is A_0-bounded with A_0-bound less than 1, then its spectrum cannot suddenly expand. In particular, we are interested in the case when the spectrum $\sigma(A_0)$ consists only of countably many isolated eigenvalues or isolated parts. If, in this case, the linear operator A_0 is self-adjoint in a Hilbert and in a Krein space, then our main results state conditions which guarantee that also the spectrum of $A_0 + V$ is real, even when $A_0 + V$ is not self-adjoint in a Hilbert space.

1.3.1 Continuity of Resolvents

In this paragraph we provide the results that will be the main tools in the following sections. While the "distance" between two bounded linear operators can be defined as the norm of their difference, the "distance" between two unbounded linear operators has to be measured in a different way. One possibility is to use the norm of the difference of their resolvents. This leads to the notion of convergence in the generalized sense considered below.

To define convergence in the generalized sense, it is necessary to consider the gap between two closed linear operators, compare [Kat95, Paragraph IV.2].

Definition 1.25. Let $\mathcal{M}$ and $\mathcal{N}$ be closed subspaces of a Banach space $\mathcal{X}$. If $\mathcal{M} \neq \{0\}$, set

$$\delta(\mathcal{M}, \mathcal{N}) := \sup_{x \in \mathcal{M}, \, \|x\| = 1} \operatorname{dist}(x, \mathcal{N}),$$
$$\widehat{\delta}(\mathcal{M}, \mathcal{N}) := \max\{\delta(\mathcal{M}, \mathcal{N}), \delta(\mathcal{N}, \mathcal{M})\};$$

if $\mathcal{M} = \{0\}$, we define $\delta(\{0\}, \mathcal{N}) = 0$ for every $\mathcal{N}$. Then $\widehat{\delta}(\mathcal{M}, \mathcal{N})$ is called the **gap** between $\mathcal{M}$ and $\mathcal{N}$.

Definition 1.26. Let $\mathcal{X}$ and $\mathcal{Y}$ be Banach spaces and let A_0 and A_1 be closed linear operators from $\mathcal{X}$ to $\mathcal{Y}$. Since A_0 and A_1 are closed, their graphs $\mathcal{G}(A)$ and $\mathcal{G}(A_1)$ are closed linear subspaces of $\mathcal{X} \times \mathcal{Y}$ and we can define

$$\delta(A_0, A_1) := \delta\big(\mathcal{G}(A_0), \mathcal{G}(A_1)\big),$$
$$\widehat{\delta}(A_0, A_1) := \widehat{\delta}\big(\mathcal{G}(A_0), \mathcal{G}(A_1)\big) = \max\{\delta(A_0, A_1), \delta(A_1, A_0)\}.$$

Then $\widehat{\delta}(A_0, A_1)$ is called the **gap** between A_0 and A_1.

If A_ε, $\varepsilon \in [0, 1]$, is a family of closed linear operators from $\mathcal{X}$ to $\mathcal{Y}$, then A_ε is said to **converge to A_0 in the generalized sense** if $\widehat{\delta}(A_\varepsilon, A_0) \to 0$ for $\varepsilon \to 0$.

For the proof of the following theorem we refer the reader to [Kat95, Theorem IV.2.25].

Theorem 1.27. *Let $\mathcal{X}$ be a Banach space and let A_ε, $\varepsilon \in [0, 1]$, be a family of closed linear operators in $\mathcal{X}$ such that $\rho(A_0) \neq \emptyset$. In order that A_ε converges to A_0 in the generalized sense, it is necessary that there exists some $\varepsilon_0 \in [0, 1]$ such that each $z \in \rho(A_0)$ also belongs to $\rho(A_\varepsilon)$ for $\varepsilon \in [0, \varepsilon_0]$ and*

$$(1.14) \qquad \left\| (A_\varepsilon - z)^{-1} - (A_0 - z)^{-1} \right\| \longrightarrow 0, \quad \varepsilon \to 0,$$

while it is sufficient that this is true for some $z \in \rho(A_0)$.

Corollary 1.28. *Theorem 1.27 implies that if (1.14) holds for one $z_0 \in \rho(A_0)$, then (1.14) is true for all $z \in \rho(A_0)$.*

Proposition 1.29. *Let A and V be closed linear operators in a Banach space $\mathcal{X}$ such that $\mathcal{D}(A) \subset \mathcal{D}(V)$. Suppose that V is A-bounded such that (1.7) holds* *with constants $\alpha \geq 0$ and $\beta \in [0,1)$. Define the family of operators $A_\varepsilon := A + \varepsilon V$, $0 \leq \varepsilon \leq 1$. If there exists a point $z \in \rho(A_0)$ such that*

$$(1.15) \qquad \alpha \|(A_0 - z)^{-1}\| + \beta \|A_0(A_0 - z)^{-1}\| < 1,$$

then $z \in \rho(A_\varepsilon)$ and A_ε converges to A_0 in the generalized sense.

Proof. By (1.15), we have

$$\|V(A_0 - z)^{-1}\| \leq \alpha \|(A_0 - z)^{-1}\| + \beta \|A_0(A_0 - z)^{-1}\| < 1.$$

and thus, using Neumann's series, we conclude $I + \varepsilon V(A_0 - z)^{-1}, 0 \leq \varepsilon \leq 1$, has a bounded inverse with domain $\mathscr{X}$. Thus also

$$A_\varepsilon - z = A_0 - z + \varepsilon V = \left(I + \varepsilon V(A_0 - z)^{-1}\right)(A_0 - z)$$

has a bounded inverse with domain $\mathscr{X}$, and hence $z \in \rho(A_\varepsilon)$ for $0 \leq \varepsilon \leq 1$. By inequality (1.15) and since V is A-bounded with A-bound less than 1, we have

$$\|\varepsilon V(A_0 - z)^{-1}x\| \leq \varepsilon \left(\alpha \|(A_0 - z)^{-1}x\| + \beta \|A_0(A_0 - z)^{-1}x\|\right)$$

for $x \in \mathscr{K}$ and $0 \leq \varepsilon \leq 1$. Since $(A_0 - z)^{-1}$ and $A_0(A_0 - z)^{-1}$ are bounded, the operator $\varepsilon V(A_0 - z)^{-1}$ is bounded with

$$\|\varepsilon V(A_0 - z)^{-1}\| \leq \varepsilon \left(\alpha \|(A_0 - z)^{-1}\| + \beta \|A_0(A_0 - z)^{-1}\|\right) < \varepsilon \longrightarrow 0, \quad \varepsilon \to 0;$$

in particular, $\|\varepsilon V(A_0 - z)^{-1}\| < 1$ for $0 \leq \varepsilon \leq 1$. Using Neumann's series, we conclude $I + \varepsilon V(A_0 - z)^{-1}$ has a bounded inverse and

$$\left\|\left(I + \varepsilon V(A_0 - z)^{-1}\right)^{-1}\right\| \leq \frac{1}{1 - \|\varepsilon V(A_0 - z)^{-1}\|}.$$

Hence $A_\varepsilon - z = \left(I + \varepsilon V(A_0 - z)^{-1}\right)(A_0 - z)$ has a bounded inverse. Using the second resolvent equation, we obtain

$$\|(A_\varepsilon - z)^{-1} - (A_0 - z)^{-1}\| \leq \|(A_\varepsilon - z)^{-1}\varepsilon V(A_0 - z)^{-1}\|$$

$$\leq \|(A_0 - z)^{-1}\| \frac{\varepsilon \|V(A_0 - z)^{-1}\|}{1 - \varepsilon \|V(A_0 - z)^{-1}\|} \longrightarrow 0$$

if $\varepsilon \to 0$. ∎

1.3.2 Perturbation of Isolated Parts of the Spectrum

In this paragraph we consider the situation where the spectrum $\sigma(A)$ of a closed linear operator A in a Banach space $\mathscr{X}$ contains a bounded subset σ_1 that is separated from the rest $\sigma(A)\backslash\sigma_1$ by a rectifiable closed curve Γ. For the case in which A is perturbed by an A-bounded operator V, we state conditions for the spectrum $A + V$ to be likewise separated. We start with an introduction to the required notation.

Definition 1.30. A **closed Jordan curve** is an oriented simple closed rectifiable curve $\gamma : [a,b] \to \mathbb{C}$, $[a,b] \subset \mathbb{R}$. A subset $\Delta \in \mathbb{C}$ is called **Cauchy domain** if there exist $\Delta_i \subset \mathbb{C}$, $i = 1,\ldots,N$, such that

   (i) $\Delta_i \neq \emptyset$, Δ_i open and connected for $i = 1,\ldots,N$,
   (ii) $\overline{\Delta_i} \cap \overline{\Delta_j} = \emptyset$, $i,j = 1,\ldots,N$, $i \neq j$,
   (iii) the boundary $\partial\Delta_i$ of Δ_i is a closed Jordan curve, $i = 1,\ldots,N$.

$\Gamma \subset \mathbb{C}$ is called **Cauchy contour** if Γ is the oriented boundary of a Cauchy domain Δ, i.e., $\Gamma = \partial\Delta$.

Definition 1.31. Let A be a closed linear operator in a Banach space $\mathscr{X}$. Then a bounded subset $\sigma \subset \sigma(A)$ is called **isolated part of** $\sigma(A)$ if σ and $\sigma(A)\backslash\sigma$ are closed. If Γ is a closed Cauchy contour such that $\sigma \subset \operatorname{int}\Gamma$ and $(\sigma(A)\backslash\sigma) \cap (\Gamma \cup \operatorname{int}\Gamma) = \emptyset$, then $\sigma(A)$ is said to be **separated into two parts** σ_1 **and** σ_2 **by** Γ.

Definition 1.32. Let A be a closed linear operator in a Banach space $\mathscr{X}$. Let $\mathscr{M}_1$ and $\mathscr{M}_2$ be two subspaces of $\mathscr{X}$. Then A is said to be **decomposed according to** $\mathscr{X} = \mathscr{M}_1 \oplus \mathscr{M}_2$ (compare [Kat95, Paragraph III.5.6]) if there exists a projection P onto $\mathscr{M}_1$ along $\mathscr{M}_2$ such that

$$P\mathscr{D}(A) \subset \mathscr{D}(A), \quad A\mathscr{M}_1 \subset \mathscr{M}_1, \quad A\mathscr{M}_2 \subset \mathscr{M}_2.$$

In this case we define the closed linear operator $A_{\mathscr{M}_1}$ in $\mathscr{M}_1$ by $\mathscr{D}(A_{\mathscr{M}_1}) := \mathscr{D}(A) \cap \mathscr{M}_1$ such that $A_{\mathscr{M}_1} x := Ax \in \mathscr{M}_1$ for $x \in \mathscr{D}(A_{\mathscr{M}_1})$. $A_{\mathscr{M}_2}$ is defined accordingly.

A proof of the following lemma can be found in [Kat95, Theorem III.6.17].

Lemma 1.33. *Let A be a closed linear operator in a Banach space $\mathscr{X}$ and let its spectrum be separated into two parts σ_1 and σ_2 by a Cauchy contour Γ.*

Then A can be decomposed according to $\mathcal{X} = \mathcal{M}_1 \oplus \mathcal{M}_2$ such that $\sigma(A_{\mathcal{M}_i}) = \sigma_i$, $i = 1, 2$, and $A_{\mathcal{M}_1}$ is bounded in $\mathcal{M}_1$. The projection P onto $\mathcal{M}_1 = P\mathcal{X}$ along $\mathcal{M}_2 = (1 - P)\mathcal{X}$ is given by

$$(1.16) \qquad P := E(A, \sigma_1) = -\frac{1}{2\pi i} \int_\Gamma (A - z)^{-1} dz.$$

Remark 1.34. The projection $E(A, \sigma_1)$ from (1.16) is called **Riesz projection** of A corresponding to σ_1 (a detailed study of Riesz projections can be found in [GGK90]).

Regarding the change of the spectrum under "small" perturbations we recall a theorem given in [Kat95], see Theorem IV.3.16 therein.

Theorem 1.35. *Let A_0 be a closed linear operator in a Banach space $\mathcal{X}$ and let the spectrum of A_0 be separated into two parts $\sigma_1(A_0)$ and $\sigma_2(A_0)$ by a Cauchy contour Γ. Let A_0 be decomposed according to $\mathcal{X} = \mathcal{M}_1(A_0) \oplus \mathcal{M}_2(A_0)$. Then there exists a $\delta > 0$, depending on A_0 and Γ, such that any closed linear operator A_1 in $\mathcal{X}$ with $\hat{\delta}(A_1, A_0) < \delta$ has spectrum $\sigma(A_1)$ likewise separated by Γ into two parts $\sigma_1(A_1)$ and $\sigma_2(A_1)$, $\Gamma \subset \rho(A_1)$. In the associated decomposition $\mathcal{X} = \mathcal{M}_1(A_1) \oplus \mathcal{M}_2(A_1)$, the subspaces $\mathcal{M}_1(A_1)$ and $\mathcal{M}_2(A_1)$ are isomorphic to $\mathcal{M}_1(A_0)$ and $\mathcal{M}_2(A_0)$, respectively. In particular,*

$$\dim \mathcal{M}_i(A_1) = \dim \mathcal{M}_i(A_0), \quad i = 1, 2,$$

and $\sigma_i(A_1)$ is non-empty if this is true for $\sigma_i(A_0)$, $i = 1, 2$. The decomposition $\mathcal{X} = \mathcal{M}_1(A_1) \oplus \mathcal{M}_2(A_1)$ is continuous in A_1 in the sense that

$$\left\| E(A_1, \sigma_1(A_1)) - E(A_0, \sigma_1(A_0)) \right\| \longrightarrow 0 \quad \text{if} \quad \hat{\delta}(A_1, A_0) \to 0.$$

Proof. According to [Kat95, Theorem IV.3.1], $\Gamma \subset \rho(A_1)$ for every closed linear operator A_1 in $\mathcal{X}$ such that $\hat{\delta}(A_1, A_0) < \delta$ with

$$\delta = \frac{1}{2} \min_{z \in \Gamma} \left((1 + |z|^2)^{-1} \left(1 + \left\| (A_0 - z)^{-1} \right\|^2 \right)^{-1/2} \right).$$

Hence $\sigma(A_1)$ is separated by Γ into two parts $\sigma_1(A_1)$ and $\sigma_2(A_1)$. By Lemma 1.33, A_1 can be decomposed according to $\mathcal{X} = \mathcal{M}_1(A_1) \oplus \mathcal{M}_2(A_1)$. We have $\mathcal{M}_1(A_1) = P\mathcal{X}$ and $\mathcal{M}_2(A_1) = (1 - P)\mathcal{X}$, where $P = E(A_1, \sigma_1(A_1))$ is the Riesz projection of A_1 corresponding to $\sigma_1(A_1)$. According to [Kat95, Theorem IV.3.15], for any $z_0 \in \rho(A_0)$ and $\varepsilon > 0$ there exists a $\delta_0 > 0$ such that

$z \in \rho(A_1)$ and $\left\| (A_1 - z)^{-1} - (A_0 - z_0)^{-1} \right\| < \varepsilon$ if $|z - z_0| < \delta_0$ and $\widehat{\delta}(A_1, A_0) < \delta_0$. Further, since Γ is compact, $\left\| (A_1 - z)^{-1} - (A_0 - z)^{-1} \right\|$ is uniformly small for $z \in \Gamma$ if $\widehat{\delta}(A_1, A_0)$ is sufficiently small. Hence

$$\left\| E(A_1, \sigma_1(A_1)) - E(A_0, \sigma_1(A_0)) \right\| \leq \frac{1}{2\pi} \int_\Gamma \left\| (A_1 - z)^{-1} - (A_0 - z)^{-1} \right\| dz$$

$$\leq \frac{l(\Gamma)}{2\pi} \left\| (A_1 - z)^{-1} - (A_0 - z)^{-1} \right\| \longrightarrow 0$$

if $\widehat{\delta}(A_1, A_0) \to 0$, where $l(\Gamma)$ is the length of the Cauchy contour Γ. The isomorphism of $\mathcal{M}_i(A_1)$ with $\mathcal{M}_i(A_0)$, $i = 1, 2$, and the fact that $\dim \mathcal{M}_i(A_1) = \dim \mathcal{M}_i(A_0)$, $i = 1, 2$, now follow from [Kat95, Lemma I.4.10] which also holds in infinite dimensional Banach spaces. ∎

The following two corollaries are special cases of Theorem 1.35 for relatively bounded perturbations.

Corollary 1.36. *Let A_0 be a closed linear operator in a Banach space $\mathcal{X}$ and let the spectrum of A_0 be separated into two parts $\sigma_1(A_0)$ and $\sigma_2(A_0)$ by a Cauchy contour Γ. Let A_0 be decomposed according to $\mathcal{X} = \mathcal{M}_1(A_0) \oplus \mathcal{M}_2(A_0)$. Further, let V be an A_0-bounded operator in $\mathcal{X}$ and consider the family of operators $A_\varepsilon := A_0 + \varepsilon V$, $0 \leq \varepsilon \leq 1$, in $\mathcal{X}$. If there exists some $\varepsilon_0 \in [0, 1]$ such that each $z \in \rho(A_0)$ belongs to $\rho(A_\varepsilon)$ for $\varepsilon \in [0, \varepsilon_0]$ and*

$$\left\| (A_\varepsilon - z)^{-1} - (A_0 - z)^{-1} \right\| \longrightarrow 0, \quad \varepsilon \to 0,$$

holds for some $z \in \rho(A_0)$, then there exists some $\varepsilon_0' \in [0, \varepsilon_0]$ such that $\sigma(A_\varepsilon)$ is likewise separated by Γ into two parts $\sigma_1(A_\varepsilon)$ and $\sigma_2(A_\varepsilon)$, $\Gamma \subset \rho(A_\varepsilon)$ and the results of Theorem 1.35 hold for $0 \leq \varepsilon \leq \varepsilon_0'$.

Proof. By Theorem 1.27, for $0 \leq \varepsilon \leq \varepsilon_0$, A_ε converges to A_0 in the generalized sense, that is, $\widehat{\delta}(A_\varepsilon, A_0) \longrightarrow 0$ for $\varepsilon \to 0$. Consequently, for any $\delta > 0$ there exists an $\varepsilon_0' \in [0, \varepsilon_0]$ such that $\widehat{\delta}(A_\varepsilon, A_0) < \delta$ if $0 \leq \varepsilon \leq \varepsilon_0'$ and thus the assumptions of Theorem 1.35 are satisfied. ∎

The following corollary can be found in [Kat95, Theorem IV.3.18].

Corollary 1.37. *Let A_0, V be defined as in the preceding corollary and consider the operator $A_1 := A_0 + V$ in $\mathcal{X}$. If*

(1.17)
$$\sup_{z \in \Gamma} \left(\alpha \left\| (A_0 - z)^{-1} \right\| + \beta \left\| A_0 (A_0 - z)^{-1} \right\| \right) < 1,$$

where α and β are constants according to (1.7), then the spectrum of A_1 is likewise separated into two parts $\sigma_1(A_1)$ and $\sigma_2(A_1)$, $\Gamma \subset \rho(A_1)$ and the results of Theorem 1.35 hold.

Proof. The proof is similar to the proof of Theorem 1.35. According to Proposition 1.29, inequality (1.17) guarantees that $\Gamma \subset \rho(A_1)$. By Lemma 1.33, A_1 can be decomposed according to $\mathscr{X} = \mathscr{M}_1(A_1) \oplus \mathscr{M}_2(A_1)$. Let $A_\varepsilon := A_0 + \varepsilon V$, $0 \leq \varepsilon \leq 1$. As in the proof of Proposition 1.29, inequality (1.17) implies that $(A_\varepsilon - z)^{-1}$, $z \in \Gamma$, depends continuously on ε for $0 \leq \varepsilon \leq 1$. Thus the Riesz projection of A_ε corresponding to $\sigma_1(A_\varepsilon)$ is continuous in ε for $0 \leq \varepsilon \leq 1$. The last part of the proof is analogous to that of Theorem 1.35.   ∎

1.3.3 Perturbation of Spectra of Self-Adjoint Operators in Hilbert Spaces

Compared to the preceding results, one can say much more about the stability of isolated parts of the spectrum if the unperturbed operator A_0 is self-adjoint in a Hilbert space. The reason for this is that the norm of the resolvent of a self-adjoint operator in a Hilbert space can be expressed in terms of the spectrum as follows. A proof of the following result may be found in [Kat95, Section V.3.8].

Proposition 1.38. *Let A be a self-adjoint operator in a Hilbert space $\mathscr{H}$. Then*

(i) $\left\| (A - z)^{-1} \right\| = \sup\limits_{\lambda \in \sigma(A)} \left(|\lambda - z|^{-1} \right), \quad z \in \rho(A),$

(ii) $\left\| A(A - z)^{-1} \right\| = \sup\limits_{\lambda \in \sigma(A)} \left(|\lambda| |\lambda - z|^{-1} \right), \quad z \in \rho(A).$

In the following, we consider families $A_\varepsilon = A_0 + \varepsilon V$, $\varepsilon \in [0,1]$, of closed linear operators where A_0 is self-adjoint in a Hilbert space $\mathscr{H}$ and V is an A_0-bounded linear operator in $\mathscr{H}$. We distinguish the following situations:

(a) We consider one isolated eigenvalue (an infinite sequence of isolated eigenvalues, respectively) of the unperturbed operator A_0.

(b) We consider one isolated compact part (an infinite sequence of isolated compact parts, respectively) of the spectrum of the unperturbed operator A_0.

Here and in the sequel for $\delta > 0$ and $z_0 \in \mathbb{C}$ or $\mathcal{M}_0 \subset \mathbb{C}$ we denote by

$$B_\delta(z_0) := \{z \in \mathbb{C} : |z - z_0| < \delta\},$$
$$B_\delta(z_0) := \{z \in \mathbb{C} : \forall z_0 \in \mathcal{M}_0 \, |z - z_0| < \delta\}$$

the δ neighbourhood of z_0 and $\mathcal{M}_0$, respectively; we also define $\mathbb{Z}^* := \{x \in \mathbb{Z} : x \neq 0\}$. In situation (a) we obtain the following result.

Theorem 1.39. *Let A_0 be a self-adjoint operator and V an A_0-bounded operator in a Hilbert space $\mathcal{H}$ with A_0-bound less than $1/2$. Define the family of operators $A_\varepsilon := A_0 + \varepsilon V$, $\varepsilon \in [0,1]$.*

(i) *Let $\lambda^0 \in \mathbb{R}$ be an isolated eigenvalue of A_0 with multiplicity $m < \infty$ and set $\delta := \mathrm{dist}(\lambda^0, \sigma(A_0) \setminus \{\lambda^0\})$. If (1.7) holds with constants $\alpha \geq 0$ and $\beta \in [0, 1/2)$ such that*

$$(1.18) \qquad \frac{1}{\delta}\Big(\alpha + \beta\big(\delta + |\lambda^0|\big)\Big) < \frac{1}{2},$$

then $\sigma(A_1) \cap B_{\delta/2}(\lambda^0)$ consists of a finite system of isolated eigenvalues with total multiplicity m.

(ii) *Let A_0 have discrete spectrum consisting of eigenvalues $\cdots < \lambda^0_{-2} < \lambda^0_{-1} < \lambda^0_1 < \lambda^0_2 < \cdots$ with multiplicities $m_n < \infty$, $n \in \mathbb{Z}^*$, and set $\delta_n := \mathrm{dist}(\lambda^0_n, \sigma(A_0) \setminus \{\lambda^0_n\})$, $n \in \mathbb{Z}^*$. If (1.7) holds with constants $\alpha_n \geq 0$ and $\beta_n \in [0, 1/2)$, $n \in \mathbb{Z}^*$, such that*

$$(1.19) \qquad \gamma := \sup_{n \in \mathbb{Z}^*}\left(\frac{1}{\delta_n}\Big(\alpha_n + \beta_n\big(\delta_n + |\lambda^0_n|\big)\Big)\right) < \infty,$$

then the spectrum of A_ε is discrete for all $\varepsilon \in [0, \varepsilon_0]$, where $\varepsilon_0 \in (0,1]$ has to be chosen such that $\varepsilon_0 < 1/(2\gamma)$. More precisely, $\sigma(A_\varepsilon) \cap B_{\delta_n/2}(\lambda^0_n)$ consists of a finite system of isolated eigenvalues with total multiplicity m_n for all $\varepsilon \in [0, \varepsilon_0]$ and $n \in \mathbb{Z}^*$.

Proof. (i). Let Γ be the positively oriented curve along the circle with center λ^0 and radius $\delta/2$. Then $\Gamma \subset \rho(A_0)$, $\{\lambda^0\} \subset \mathrm{int}\,\Gamma$ and $(\Gamma \cup \mathrm{int}\,\Gamma) \cap (\sigma(A_0) \setminus \{\lambda^0\}) = \emptyset$.

If $z \in \Gamma$, then, by the choice of Γ, $|\lambda - z|^{-1} \leq 2/\delta$, $\lambda \in \sigma(A_0)$. Furthermore, since $|\lambda^0 - z| = \delta/2$, we obtain that for $\lambda \in \sigma(A_0)$

$$|\lambda||\lambda - z|^{-1} \leq \left(|\lambda - z| + |z|\right)|\lambda - z|^{-1}$$
$$\leq 1 + \left(|z - \lambda^0| + |\lambda^0|\right)|\lambda - z|^{-1}$$
$$\leq 2 + \frac{2|\lambda^0|}{\delta}.$$

By Proposition 1.38, since A_0 is self-adjoint in the Hilbert space $\mathcal{H}$, inequality (1.17) is satisfied if

$$\alpha \frac{2}{\delta} + \beta\left(2 + \frac{2|\lambda^0|}{\delta}\right) < 1,$$

or, equivalently, (1.18) holds. The application of Corollary 1.37 completes the proof.

(ii). By (1.19), the assumptions of part (i) are satisfied for A_0 and εV with $\varepsilon \in [0, \varepsilon_0]$ for every eigenvalue λ_n^0, $n \in \mathbb{Z}^*$, if we choose ε_0 such that

$$\varepsilon_0 \sup_{n \in \mathbb{Z}^*}\left(\frac{1}{\delta_n}\left(\alpha_n + \beta_n(\delta_n + |\lambda_n^0|)\right)\right) < \frac{1}{2}. \qquad \blacksquare$$

The following theorem deals with situation (b), where isolated compact parts of the spectrum of the unperturbed operator are considered.

Theorem 1.40. *Let A_0 be a self-adjoint operator and V an A_0-bounded operator in a Hilbert space $\mathcal{H}$ with A_0-bound less than $1/2$. Define the family of operators $A_\varepsilon := A_0 + \varepsilon V$, $\varepsilon \in [0, 1]$.*

(i) *Let σ_0 be an isolated part of $\sigma(A_0)$ such that $\sigma_0 = \sigma(A_0) \cap [\lambda^-, \lambda^+]$ with $\lambda^-, \lambda^+ \in \mathbb{R}$ and set $\delta := \mathrm{dist}(\sigma_0, \sigma(A_0)\backslash\sigma_0)$. If (1.7) holds with constants $\alpha \geq 0$ and $\beta \in [0, 1/2)$ such that*

(1.20)
$$\frac{1}{\delta}\left(\alpha + \beta\left(\delta + \max\left\{|\lambda^-|, |\lambda^+|\right\}\right)\right) < \frac{1}{2},$$

then $\partial B_{\delta/2}([\lambda^-, \lambda^+]) \subset \rho(A_\varepsilon)$ and $\sigma_\varepsilon := \sigma(A_\varepsilon) \cap B_{\delta/2}([\lambda^-, \lambda^+])$ is an isolated part of $\sigma(A_\varepsilon)$ for all $\varepsilon \in [0, 1]$. Furthermore, $\dim E(A_0, \sigma_0)\mathcal{K} = \dim E(A_\varepsilon, \sigma_\varepsilon)\mathcal{K}$ for $\varepsilon \in [0, 1]$.

(ii) *Let*

$$\sigma(A_0) = \bigcup_{n \in \mathbb{Z}^*} \sigma_{n,0}$$

with $\sigma_{n,0} = \sigma(A_0) \cap [\lambda_n^-, \lambda_n^+]$, $n \in \mathbb{Z}^*$, *where* $\lambda_n^-, \lambda_n^+ \in \mathbb{R}$, *and*

$$\cdots < \lambda_{-2}^- \le \lambda_{-2}^+ < \lambda_{-1}^- \le \lambda_{-1}^+ < \lambda_1^- \le \lambda_1^+ < \lambda_2^- \le \lambda_2^+ < \cdots.$$

Define $\delta_n := \mathrm{dist}(\sigma_{n,0}, \sigma(A_0) \setminus \sigma_{n,0})$, $n \in \mathbb{Z}^*$. *If* (1.7) *holds with constants* $\alpha_n \ge 0$ *and* $\beta_n \in [0, 1/2)$, $n \in \mathbb{Z}^*$, *such that*

(1.21) $$\gamma := \sup_{n \in \mathbb{Z}^*} \left(\frac{1}{\delta_n} \left(\alpha_n + \beta_n \left(\delta_n + \max\{|\lambda_n^-|, |\lambda_n^+|\} \right) \right) \right) < \infty,$$

then $\partial B_{\delta_n/2}([\lambda_n^-, \lambda_n^+]) \subset \rho(A_\varepsilon)$ *and* $\sigma_{n,\varepsilon} := \sigma(A_\varepsilon) \cap B_{\delta_n/2}([\lambda_n^-, \lambda_n^+])$ *is an isolated part of* $\sigma(A_\varepsilon)$ *for all* $n \in \mathbb{Z}^*$, $\varepsilon \in [0, \varepsilon_0]$, *and*

$$\sigma(A_\varepsilon) = \bigcup_{n \in \mathbb{Z}^*} \sigma_{n,\varepsilon}, \quad \varepsilon \in [0, \varepsilon_0],$$

where $\varepsilon_0 \in (0, 1]$ *has to be chosen such that* $\varepsilon_0 < 1/(2\gamma)$. *Furthermore,* $\dim E(A_0, \sigma_{n,0})\mathcal{K} = \dim E(A_\varepsilon, \sigma_{n,\varepsilon})\mathcal{K}$ *for* $\varepsilon \in [0, \varepsilon_0]$ *and* $n \in \mathbb{Z}^*$.

Proof. (i). Let $\Gamma \subset \rho(A_0)$ be the positively oriented curve along $\partial B_{\delta/2}([\lambda^-, \lambda^+])$. For all $z \in \Gamma$ there exists a $\lambda_z \in [\lambda_0^-, \lambda_0^+]$ such that $|\lambda - z|^{-1} \le |\lambda_z - z|^{-1} = \delta/2$, $\lambda \in \sigma(A_0)$. For any $z \in \Gamma$ and any $\lambda \in \sigma(A_0)$ we have, as above,

$$|\lambda||\lambda - z|^{-1} \le 2 + \frac{2|\lambda_z|}{\delta} \le 2 + \frac{2\max\{|\lambda_n^-|, |\lambda_n^+|\}}{\delta}.$$

Since inequality (1.20) satisfies inequality (1.17), the statement follows from Corollary 1.37.

(ii). By (1.21), the assumptions of part (i) are satisfied for A_0 and εV with $\varepsilon \in [0, \varepsilon_0]$ for every isolated compact part $\sigma_{n,0}$, $n \in \mathbb{Z}^*$, if we choose ε_0 such that

$$\varepsilon_0 \sup_{n \in \mathbb{Z}^*} \left(\frac{1}{\delta_n} \left(\alpha_n + \beta_n \left(\delta_n + \max\{|\lambda_n^-|, |\lambda_n^+|\} \right) \right) \right) < \frac{1}{2}. \qquad \blacksquare$$

Remark 1.41. If in Theorem 1.39 (ii) and Theorem 1.40 (ii) A_0 is semibounded, $\lambda_1^0 < \lambda_2^0 < \cdots$ or $\lambda_1^- \le \lambda_1^+ < \lambda_2^- \le \lambda_2^+ < \cdots$, it follows that $\sigma(A_0 + \varepsilon V)$ is bounded from below for $\varepsilon \in [0, \varepsilon_0]$.

Remark 1.42. Applying similar arguments as in Theorem 1.40 for the case of one spectral gap (a, b) in the spectrum of a self-adjoint operator A_0, one may obtain the following result, compare [Cue]. Let V be A_0-bounded with A_0-bound less than 1. If

$$(1.22) \qquad \frac{1}{\delta}\Big(\alpha + \beta \max\{|a|, |b|\}\Big) < \frac{1}{2},$$

where α and β are constants according to (1.7) and $\delta := b - a$, then the open strip $I + i\mathbb{R}$, where the interval I is given by

$$I := \Big(a + \big(\alpha + \beta \max\{|a|, |b|\}\big),\ b - \big(\alpha + \beta \max\{|a|, |b|\}\big)\Big),$$

is contained in the resolvent set $\rho(A_1)$ of $A_1 := A_0 + V$. Note that the interval I is non-empty if inequality (1.22) is satisfied.

As in the proof of Theorem 1.40 one can show that

$$\alpha \big\|(A_0 - z)^{-1}\big\| + \beta \big\|A_0 (A_0 - z)^{-1}\big\| < 1$$

if $z \in I$, and hence $z \in \rho(A_1)$ by [Kat95, Theorem IV.3.17].

1.3.4 Perturbation of Spectra of Self-Adjoint Operators in Krein Spaces

In this section we consider the case where A_0 and V are self-adjoint and symmetric in a Krein space, respectively. We combine Theorems 1.39 and 1.40 on the stability of isolated parts of the spectrum with a stability result for uniformly positive subspaces of a Krein space. This allows us to establish conditions guaranteeing that the spectrum of $A_0 + V$ is real provided the spectrum of A_0 is real.

The following stability result for uniformly definite subspaces of a Krein space was proved in [LT04, Theorem 3.1]. This theorem has already been applied for bounded perturbations V in [LT04, Theorem 3.1] and it is also fundamental for unbounded perturbations considered below.

Theorem 1.43. Let A_ε, $0 \le \varepsilon \le 1$, be a family of self-adjoint operators in a Krein space $\big(\mathscr{K}, [\cdot, \cdot]\big)$ such that for one (and hence for all) $z \in \rho(A_\varepsilon)$ the resolvents $(A_\varepsilon - z)^{-1}$ depend continuously on ε in the operator norm for $0 \le \varepsilon \le 1$. Let σ_0 be an isolated part of the spectrum $\sigma(A_0)$ of A_0 such that $\sigma_0 = \sigma_0^*$, and let Γ be a Cauchy contour surrounding σ_0 such that $\Gamma = \Gamma^*$. Suppose *that* $\sigma(A_\varepsilon) \cap \Gamma = \emptyset$ *for all* $0 \le \varepsilon \le 1$ *and set* $\sigma_\varepsilon := \sigma(A_\varepsilon) \cap \operatorname{int}\Gamma$. *If the subspace* $E(A_0, \sigma_0)\mathscr{K}$ *is uniformly positive (uniformly negative, respectively), then for all* $0 \le \varepsilon \le 1$ *the subspaces* $E(A_\varepsilon, \sigma_\varepsilon)\mathscr{K}$ *are uniformly positive (uniformly negative, respectively) and the sets* σ_ε *are real.*

As in the preceding paragraph the following two theorems cover situations (a) (isolated eigenvalues) and (b) (isolated compact parts of the spectrum), respectively.

Theorem 1.44. *Let* A_0 *be a self-adjoint operator in a Krein space* $(\mathscr{K}, [\cdot, \cdot])$ *such that* A_0 *is also self-adjoint in the Hilbert space* $(\mathscr{K}, (\cdot, \cdot))$. *Further, let* V *be a symmetric operator in the Krein space* $(\mathscr{K}, [\cdot, \cdot])$ *which is* A_0-*bounded with* A_0-*bound less than* $1/2$ *and define the family of operators* $A_\varepsilon := A_0 + \varepsilon V$, $\varepsilon \in [0, 1]$.

 (i) *Let* $\lambda^0 \in \mathbb{R}$ *be an isolated eigenvalue of* A_0 *which is of definite type with multiplicity* $m < \infty$ *and set* $\delta := \operatorname{dist}(\lambda^0, \sigma(A_0) \setminus \{\lambda^0\})$. *If* (1.7) *holds with constants* $\alpha \ge 0$ *and* $\beta \in [0, 1/2)$ *such that*

$$(1.23) \qquad \frac{1}{\delta}\Big(\alpha + \beta\big(\delta + |\lambda^0|\big)\Big) < \frac{1}{2},$$

 then $\sigma(A_1) \cap B_{\delta/2}(\lambda^0)$ *consists of a finite system of isolated and real eigenvalues with total multiplicity* m *which are of the same type as* λ^0.

 (ii) *Let* A_0 *have discrete spectrum consisting of eigenvalues* $\cdots < \lambda_{-2}^0 < \lambda_{-1}^0 < \lambda_1^0 < \lambda_2^0 < \cdots$ *of definite type with multiplicities* $m_n < \infty$, $n \in \mathbb{Z}^*$, *and define* $\delta_n := \operatorname{dist}(\lambda_n^0, \sigma(A_0) \setminus \{\lambda_n^0\})$, $n \in \mathbb{Z}^*$. *If* (1.7) *holds with constants* $\alpha_n \ge 0$ *and* $\beta_n \in [0, 1/2)$, $n \in \mathbb{Z}^*$, *such that*

$$(1.24) \qquad \gamma := \sup_{n \in \mathbb{Z}^*}\left(\frac{1}{\delta_n}\Big(\alpha_n + \beta_n\big(\delta_n + |\lambda_n^0|\big)\Big)\right) < \infty,$$

 then the spectrum of A_ε *is discrete and consists of real eigenvalues of definite type for all* $\varepsilon \in [0, \varepsilon_0]$, *where* $\varepsilon_0 \in (0, 1]$ *has to be chosen such that* $\varepsilon_0 < 1/(2\gamma)$. *More precisely,* $\sigma(A_\varepsilon) \cap B_{\delta_n/2}(\lambda_n^0)$ *consists of a finite system of real isolated eigenvalues with total multiplicity* m_n *which are of the same type as* λ_n^0 *for all* $\varepsilon \in [0, \varepsilon_0]$ *and* $n \in \mathbb{Z}^*$.

Proof. (i). For the proof we apply Corollary 1.37 and Theorem 1.43. The family of operators A_ε, $0 \le \varepsilon \le 1$, is self-adjoint in the Krein space $(\mathscr{K}, [\cdot, \cdot])$ by Theorem 1.24. Let Γ be the positively oriented curve along the circle with center λ^0 and radius $\delta/2$. Then $\Gamma \subset \rho(A_0)$ separates the spectrum of A_0 into the two parts $\{\lambda^0\}$ and $\sigma(A_0)\backslash\{\lambda^0\}$. As we have seen in the proof of Theorem 1.39, by inequality (1.23), the assumptions of Corollary 1.37 are satisfied. Thus the spectrum of $A_\varepsilon = A_0 + \varepsilon V$ is also separated into two parts by Γ, where Γ itself is running in $\rho(A_\varepsilon)$. According to Corollary 1.37, the eigenvalue λ^0 of A_0 splits at most into a finite system of isolated eigenvalues of A_ε enclosed by Γ which have total multiplicity m. By Proposition 1.29, A_ε converges to A_0 in the generalized sense. Now Theorem 1.43 can be applied. If we define $\sigma_\varepsilon := \sigma(A_\varepsilon) \cap \text{int}\,\Gamma$, $0 \leq \varepsilon \leq 1$, then $\sigma_\varepsilon = \sigma_\varepsilon^*$ for all $0 \leq \varepsilon \leq 1$. This is due to the fact that spectrum of the operator A_ε, which is self-adjoint in $(\mathscr{K},[\cdot,\cdot])$, is symmetric with respect to the real axis and that $\Gamma = \Gamma^*$. We note that $\sigma_0 = \{\lambda^0\}$. Since, by assumption, λ^0 is of definite type, $E(A_0,\sigma_0)\mathscr{K}$ is either uniformly positive or uniformly negative. According to Theorem 1.43 the subspace $E(A_\varepsilon,\sigma_\varepsilon)\mathscr{K}$ is of the same type as $E(A_0,\sigma_0)\mathscr{K}$ and the set σ_ε is real for all $0 \leq \varepsilon \leq 1$. Hence σ_ε consists of a finite system of real eigenvalues of A_ε with total multiplicity m and which is of the same type as λ^0.

(ii). By (1.24), the assumptions of part (i) are satisfied for A_0 and εV with $\varepsilon \in [0,\varepsilon_0]$ for every eigenvalue λ_n^0, $n \in \mathbb{Z}^*$, if we choose ε_0 such that

$$\varepsilon_0 \sup_{n \in \mathbb{Z}^*}\left(\frac{1}{\delta_n}\left(\alpha_n + \beta_n\left(\delta_n + |\lambda_n^0|\right)\right)\right) < \frac{1}{2}. \qquad \blacksquare$$

Remark 1.45. For a special class of differential operators the last corollary has also been proved in a paper by E. Caliceti and S. Graffi (see [CG05]; see also [Cal05] and [CCG06]), compare Theorem 3.20 and Remark 3.21 below. They consider operators induced by the differential expression

$$A_\varepsilon = -\frac{\mathrm{d}^2}{\mathrm{d}x^2} + P + \varepsilon \mathrm{i} Q, \quad \varepsilon \in [0,1],$$

in $L^2(\mathbb{R})$, where P and Q are multiplication operators by real polynomials P and Q. In Section 3.3 we show that the conditions of Theorem 1.44 (ii) are satisfied in this special case with

$$A_0 = -\frac{\mathrm{d}^2}{\mathrm{d}x^2} + P \quad \text{and} \quad V = \mathrm{i} Q.$$

Theorem 1.46. *Let A_0 be a self-adjoint operator in a Krein space $(\mathscr{K},[\cdot,\cdot])$ such that A_0 is also self-adjoint in the Hilbert space $(\mathscr{K},(\cdot,\cdot))$. Further, let V be a symmetric operator in the Krein space $(\mathscr{K},[\cdot,\cdot])$ which is A_0-bounded* *with A_0-bound less than $1/2$ and define the family of operators $A_\varepsilon := A_0 + \varepsilon V$, $\varepsilon \in [0,1]$.*

(i) *Let σ_0 be an isolated part of $\sigma(A_0)$ such that $E(A_0,\sigma_0)\mathscr{K}$ is uniformly definite with $\sigma_0 = \sigma(A_0) \cap [\lambda^-, \lambda^+]$, where $\lambda^-, \lambda^+ \in \mathbb{R}$. Define the distance $\delta := \mathrm{dist}(\sigma_0, \sigma(A_0)\backslash\sigma_0)$. If (1.7) holds with constants $\alpha \geq 0$ and $\beta \in [0,1/2)$ such that*

$$\frac{1}{\delta}\left(\alpha + \beta\left(\delta + \max\{|\lambda^-|, |\lambda^+|\}\right)\right) < \frac{1}{2},$$

then the set $\sigma_\varepsilon := \sigma(A_\varepsilon) \cap B_{\delta/2}([\lambda^-, \lambda^+])$ is real and an isolated part of $\sigma(A_\varepsilon)$ for all $\varepsilon \in [0,1]$. Furthermore, $\dim E(A_0,\sigma_0)\mathscr{K} = \dim E(A_\varepsilon,\sigma_\varepsilon)\mathscr{K}$ for $\varepsilon \in [0,1]$.

(ii) *Let*

$$\sigma(A_0) = \bigcup_{n\in\mathbb{Z}^*} \sigma_{n,0}$$

with $\sigma_{n,0} = \sigma(A_0) \cap [\lambda_n^-, \lambda_n^+]$, $n \in \mathbb{Z}^*$, where $\lambda_n^-, \lambda_n^+ \in \mathbb{R}$, $n \in \mathbb{Z}^*$, and

$$\cdots < \lambda_{-2}^- \leq \lambda_{-2}^+ < \lambda_{-1}^- \leq \lambda_{-1}^+ < \lambda_1^- \leq \lambda_1^+ < \lambda_2^- \leq \lambda_2^+ < \cdots.$$

Suppose that $E(A_0,\sigma_{n,0})\mathscr{K}$ is uniformly definite for $n \in \mathbb{Z}^*$. Define the distances $\delta_n := \mathrm{dist}(\sigma_{n,0}, \sigma(A_0)\backslash\sigma_{n,0})$, $n \in \mathbb{Z}^*$. If (1.7) holds with constants $\alpha_n \geq 0$ and $\beta_n \in [0,1/2)$, $n \in \mathbb{Z}^*$, such that

$$\gamma := \sup_{n\in\mathbb{Z}^*}\left(\frac{1}{\delta_n}\left(\alpha_n + \beta_n\left(\delta_n + \max\{|\lambda_n^-|, |\lambda_n^+|\}\right)\right)\right) < \infty,$$

then the sets $\sigma_{n,\varepsilon} := \sigma(A_\varepsilon) \cap B_{\delta_n/2}([\lambda_n^-, \lambda_n^+])$ are real and isolated parts of $\sigma(A_\varepsilon)$ and

$$\sigma(A_\varepsilon) = \bigcup_{n\in\mathbb{Z}^*} \sigma_{n,\varepsilon} \subset \mathbb{R}, \quad \varepsilon \in [0,\varepsilon_0],$$

where $\varepsilon_0 \in (0,1]$ has to be chosen such that $\varepsilon_0 < 1/(2\gamma)$. Furthermore, $\dim E(A_0,\sigma_{n,0})\mathscr{K} = \dim E(A_\varepsilon,\sigma_{n,\varepsilon})\mathscr{K}$ for $\varepsilon \in [0,\varepsilon_0]$ and $n \in \mathbb{Z}^*$.

Proof. With the help Theorem 1.40, the proof is analogous to the proof of Theorem 1.44. ∎

When, in contrast to the two preceding theorems, the unperturbed operator A_0 is only self-adjoint in a Krein space $(\mathscr{K},[\cdot,\cdot])$, then we have the following result.

Theorem 1.47. *Let A_0 be a self-adjoint operator in a Krein space $(\mathcal{K}, [\cdot, \cdot])$ with discrete real spectrum consisting of eigenvalues $\cdots < \lambda_{-2}^0 < \lambda_{-1}^0 < \lambda_1^0 < \lambda_2^0 < \cdots$ of definite type. Let $\Gamma_n \subset \rho(A_0)$, $n \in \mathbb{Z}^*$, be a closed Jordan curve surrounding the eigenvalue λ_n^0 of A_0 such that $\Gamma_n = \Gamma_n^*$ and $(\Gamma_n \cup \mathrm{int}(\Gamma_n)) \cap (\sigma(A_0) \backslash \{\lambda_n^0\}) = \emptyset$. Set $\Gamma := \bigcup_{n \in \mathbb{Z}^*} \Gamma_n$. Let V be a symmetric operator in the Krein space $(\mathcal{K}, [\cdot, \cdot])$ which is A_0-bounded. If (1.7) holds with constants $\alpha \geq 0$ and $\beta \in [0, 1)$ such that*

$$(1.25) \qquad \gamma := \sup_{z \in \Gamma} \left(\alpha \|(A_0 - z)^{-1}\| + \beta \|A_0(A_0 - z)^{-1}\| \right) < \infty,$$

then the spectrum of $A_0 + \varepsilon V$ is discrete and real and consists of eigenvalues of definite type for all $\varepsilon \in [0, \varepsilon_0]$, where $\varepsilon_0 \in (0, 1]$ has to be chosen such that $\varepsilon_0 < 1/(2\gamma)$. More precisely, if m_n denotes the multiplicity of the eigenvalues λ_n^0 of A_0, $n \in \mathbb{Z}^*$, then under the perturbation εV the eigenvalues λ_n^0 of A_0 split into a finite system of isolated eigenvalues with total multiplicity m_n which are of the same type as λ_n^0.

Proof. Since A_0 is not assumed to be self-adjoint in a Hilbert space, we cannot use Theorem 1.39, but verify the assumptions of Corollary 1.37 directly. In fact, for each $n \in \mathbb{Z}^*$ the spectrum of A_0 is separated into two parts by Γ_n. If we choose ε_0 such that

$$\varepsilon_0 \sup_{z \in \Gamma} \left(\alpha \|(A_0 - z)^{-1}\| + \beta \|A_0(A_0 - z)^{-1}\| \right) < \frac{1}{2},$$

then the assumptions of Corollary 1.37 are satisfied for A_0 and εV with $\varepsilon \in [0, \varepsilon_0]$ for every λ_n^0, $n \in \mathbb{Z}^*$. Now the results follow as in the proof of Theorem 1.44. ∎

Remark 1.48. If V is bounded, then $\alpha = \|V\|$, $\beta = 0$ and [LT04, Corollary 3.4] is a special case of Theorem 1.47 (see also [LT06]).

Chapter 2

Relatively Form-Bounded Perturbations in Krein Spaces

In Chapter 1 we considered perturbations $A_0 + V$ of a closed linear operator A_0 by a linear operator V for which $A_0 + V$ is defined as an operator sum. In this chapter we consider operators A_0 and V for which the sum is defined only by means of quadratic forms. We will extend the main results from Chapter 1 to this more general case.

2.1 Stability Theorems

As in the case of relatively bounded perturbations, a number of important properties of linear operators are preserved under relatively form-bounded perturbations. We recall the basic definitions and give a short review on quadratic forms and associated operators. A detailed study of quadratic forms as well as of relatively form-bounded and relatively form-compact operators can be found in [Kat95], [RS80], [RS75], [RS78] and [EE87].

2.1.1 Accretive and Sectorial Operators

Definition 2.1. (i) Let $(\mathcal{K}, [\cdot, \cdot])$ be a Krein space and let A be a linear operator in $(\mathcal{K}, [\cdot, \cdot])$. We define the set $W^{[*]}(A) \subset \mathbb{C}$ by

$$W^{[*]}(A) := \{[Ax, x] : x \in \mathcal{D}(A), \|x\| = 1\}.$$

(ii) A symmetric operator A in $(\mathcal{K}, [\cdot, \cdot])$ is called **bounded from below** (*in the Krein space* $(\mathcal{K}, [\cdot, \cdot])$) if $W^{[*]}(A)$ (which is a subset of $\mathbb{R}$) is bounded from below, that is, if there exists some $\gamma \in \mathbb{R}$ such that

$$[Ax, x] \geq \gamma \|x\|^2, \quad x \in \mathcal{D}(A).$$

If $\gamma = 0$, then A is said to be **non-negative** (*in the Krein space* $(\mathcal{K}, [\cdot, \cdot])$).

(iii) The linear operator A is said to be **accretive** (**in the Krein space** $(\mathcal{K}, [\cdot, \cdot])$) if $W^{[*]}(A)$ is a subset of the right half-plane, i.e., if

$$\mathrm{Re}\,[Ax, x] \geq 0, \quad x \in \mathcal{D}(A).$$

Further, A is called **quasi-accretive** (**in the Krein space** $(\mathcal{K}, [\cdot, \cdot])$) if $A + \gamma$ is accretive for some $\gamma > 0$.

(iv) In the case when A is additionally closed, A is said to be **m-accretive** (**in the Krein space** $(\mathcal{K}, [\cdot, \cdot])$) if it admits no non-trivial accretive extensions. If $A + \gamma$ is m-accretive for some $\gamma > 0$, then A is called **quasi-m-accretive** (**in the Krein space** $(\mathcal{K}, [\cdot, \cdot])$).

(v) The linear operator A is called **sectorially-valued** or simply **sectorial** (**in the Krein space** $(\mathcal{K}, [\cdot, \cdot])$) if $W^{[*]}(A)$ is a subset of a sector

$$\left\{ z \in \mathbb{C} : \mathrm{Re}\,z \geq \gamma, \ |\arg(z - \gamma)| \leq \theta < \frac{\pi}{2} \right\},$$

for some $\gamma \in \mathbb{R}$ and $0 \leq \theta < \pi/2$. We shall call γ a **vertex** and θ a **semi-angle** of the sectorial operator A. Note that a sectorial operator is quasi-accretive. A is called **m-sectorial** (**in the Krein space** $(\mathcal{K}, [\cdot, \cdot])$) if it is sectorial and quasi-m-accretive.

Remark 2.2. The above notations are sometimes also referred to as J-non-negative, J-accretive etc. because, e.g., A is non-negative in the Krein space $(\mathcal{K}, [\cdot, \cdot])$ if and only if JA is non-negative in the Hilbert space $(\mathcal{K}, (\cdot, \cdot))$. If the (classical) numerical range of A in a Hilbert space is denoted by $W(A)$, then $W^{[*]}(A) = W(JA)$ and vice versa since $[x, y] = (Jx, y)$, $x, y \in \mathcal{K}$.

Remark 2.3. The set $W^{[*]}(A)$ differs from the set called Krein space numerical range which was defined in [LTU96] as $\{[Ax, x] : x \in \mathcal{D}(A), [x, x] = 1\}$.

2.1.2 Quadratic Forms and Associated Operators

Definition 2.4. Let $\mathcal{X}$ be a vector space over $\mathbb{K}$ ($\mathbb{K} = \mathbb{R}$ or $\mathbb{C}$). A **sesquilinear form** is a map $\mathfrak{a} : \mathcal{D} \times \mathcal{D} \to \mathbb{C}$ such that

(i) $\mathfrak{a}[\alpha x_1 + \beta x_2, y] = \alpha \mathfrak{a}[x_1, y] + \beta \mathfrak{a}[x_2, y]$,

(ii) $\mathfrak{a}[x, \alpha y_1 + \beta y_2] = \overline{\alpha} \mathfrak{a}[x, y_1] + \overline{\beta} \mathfrak{a}[x, y_2]$,

where $\mathcal{D}$ is a subspace of $\mathcal{X}$ and $x_1, x_2, y_1, y_2 \in \mathcal{D}$, $\alpha, \beta \in \mathbb{K}$. Further, $\mathfrak{a} : \mathcal{D} \to \mathbb{C}$, $\mathfrak{a}[x] = \mathfrak{a}[x, x]$ is called **quadratic form** associated with $\mathfrak{a}[x, y]$. The subspace $\mathcal{D} = \mathcal{D}(\mathfrak{a})$ is called domain of the sesquilinear or quadratic form $\mathfrak{a}$. When there is no possibility of confusion we call the sesquilinear form $\mathfrak{a}[\![x,y]\!]$ or quadratic form $\mathfrak{a}[\![x]\!]$ simply a **form**.

Definition 2.5.   (i) Let $\mathscr{X}$ be a vector space over $\mathbb{K}$ ($\mathbb{K} = \mathbb{R}$ or $\mathbb{C}$). The sesquilinear form $\mathfrak{a}^*$ defined by

$$\mathfrak{a}^*[\![x,y]\!] := \overline{\mathfrak{a}[\![y,x]\!]}, \quad x,y \in \mathscr{D}(\mathfrak{a}^*) = \mathscr{D}(\mathfrak{a}),$$

is called the **adjoint form** of $\mathfrak{a}$. A form $\mathfrak{a} : \mathscr{D}(\mathfrak{a}) \times \mathscr{D}(\mathfrak{a}) \to \mathbb{C}$ is said to be **symmetric** if $\mathfrak{a}^* = \mathfrak{a}$, that is, if

$$\mathfrak{a}[\![x,y]\!] = \overline{\mathfrak{a}[\![y,x]\!]}, \quad x,y \in \mathscr{D}(\mathfrak{a});$$

in this case the quadratic form $\mathfrak{a}$ is real valued.

 (ii) A symmetric form $\mathfrak{a}$ is called **bounded from below** if there exists a $\gamma \in \mathbb{R}$ such that $\mathfrak{a} \geq \gamma$, that is,

$$\mathfrak{a}[\![x]\!] \geq \gamma \|x\|^2, \quad x \in \mathscr{D}(\mathfrak{a}).$$

If $\gamma \geq 0$, then $\mathfrak{a}$ is called **non-negative**. The notions of boundedness from above and non-positiveness may be defined analogously.

(iii) For a quadratic form $\mathfrak{a}$ we define the **numerical range** $W(\mathfrak{a})$ by

$$W(\mathfrak{a}) := \big\{\mathfrak{a}[\![x]\!] : x \in \mathscr{D}(\mathfrak{a}), \|x\| = 1\big\}.$$

(iv) The quadratic form $\mathfrak{a}$ is called **sectorially bounded** from the left or simply **sectorial** if $W(\mathfrak{a})$ is a subset of a sector

$$\Big\{z \in \mathbb{C} : \operatorname{Re} z \geq \gamma, \ |\arg(z - \gamma)| \leq \theta < \frac{\pi}{2}\Big\},$$

for some $\gamma \in \mathbb{R}$ and $0 \leq \theta < \pi/2$. We call γ a **vertex** and θ a **semi-angle** of the form $\mathfrak{a}$.

In the following we define closed and closable forms. First we have to introduce the notion of $\mathfrak{a}$-convergence.

Definition 2.6.   (i) Let $\mathfrak{a}$ be a form in a Banach space $\mathscr{X}$. A sequence $(x_n)_{n=0}^{\infty} \subset \mathscr{D}(\mathfrak{a})$ is called $\mathfrak{a}$-**convergent** (**to** $x \in \mathscr{X}$) if $x_n \to x$ and $\mathfrak{a}[\![x_n - x_m]\!] \to 0$ for $m, n \to \infty$.

 (ii) If $\mathfrak{a}$ is sectorial, then $\mathfrak{a}$ is said to be **closed** if for any $\mathfrak{a}$-convergent sequence $(x_n)_{n=0}^{\infty} \subset \mathscr{D}(\mathfrak{a})$ we have $x \in \mathscr{D}(\mathfrak{a})$ and $\mathfrak{a}[\![x_n - x]\!] \to 0$ for $n \to \infty$.

(iii) A sectorial form $\mathfrak{a}$ is said to be **closable** if it has a closed extension; in this case the smallest closed extension $\bar{\mathfrak{a}}$ is called the **closure** of $\mathfrak{a}$.

Example 2.7. Let A be a linear operator in a Krein space $(\mathcal{K}, [\cdot, \cdot])$. Then A induces a sesquilinear form $\mathfrak{a}$ on $\mathcal{K}$ by means of

$$\mathfrak{a}[x, y] := [Ax, y], \quad x, y \in \mathcal{D}(\mathfrak{a}) = \mathcal{D}(A).$$

Clearly, if A is symmetric in $(\mathcal{K}, [\cdot, \cdot])$, then the form $\mathfrak{a}$ is symmetric. If A is sectorial with vertex γ, then $\mathfrak{a}$ is sectorial with vertex γ.

Definition 2.8. Let $\mathfrak{a}$ be a closed sectorial form in a Krein space $\mathcal{K}$ with vertex γ and define

$$\|x\|_{\mathfrak{a}} := (\operatorname{Re} \mathfrak{a} - \gamma + 1)^{1/2} [\![x]\!], \quad x \in \mathcal{D}(\mathfrak{a}).$$

A subspace $\mathcal{D}$ of $\mathcal{K}$ is said to be a **core** of $\mathfrak{a}$ if $\mathcal{D}$ is dense in the Banach space $(\mathcal{D}(\mathfrak{a}), \|\cdot\|_{\mathfrak{a}})$ (compare, e.g., [Kat95, Section VI.1.3]).

The following theorem is a generalization of the well-known first representation theorem for quadratic forms in a Hilbert space (see [Kat95, Paragraph VI.2]) to quadratic forms in a Krein space (see [Fle99, Theorem 1]).

Theorem 2.9 (The first representation theorem). *Let $(\mathcal{K}, [\cdot, \cdot])$ be a Krein space and let $\mathfrak{a}$ be a sectorial sesquilinear form in $\mathcal{K}$ which is closed with respect to the Hilbert space topology of $\mathcal{K}$. Then there exists an m-sectorial operator A in $(\mathcal{K}, [\cdot, \cdot])$ such that*

(i) $\mathcal{D}(A) \subset \mathcal{D}(\mathfrak{a})$ *and*

$$\mathfrak{a}[x, y] = [Ax, y], \quad x \in \mathcal{D}(A), y \in \mathcal{D}(\mathfrak{a});$$

(ii) $\mathcal{D}(A)$ *is a core of* $\mathfrak{a}$;

(iii) *if* $x \in \mathcal{D}(\mathfrak{a})$, $z \in \mathcal{K}$ *and*

$$\mathfrak{a}[x, y] = [z, y]$$

holds for every y belonging to a core of $\mathfrak{a}$, then $x \in \mathcal{D}(A)$ and $Ax = z$.

Proof. According to the Hilbert space version of the first representation theorem, see, e.g., [Kat95, Theorem VI.2.1], there exists an m-sectorial operator

$\tilde{A}_\mathfrak{a}$ in the Hilbert space $(\mathcal{K},(\cdot,\cdot))$ such that above conditions (i) to (iii) hold for $\tilde{A}_\mathfrak{a}$ with $[\cdot,\cdot]$ replaced by $(\cdot,\cdot)$ in (i) and (iii). If we define

$$(2.1) \qquad\qquad A_\mathfrak{a} := J\tilde{A}_\mathfrak{a},$$

then $A_\mathfrak{a}$ is m-sectorial in the Krein space $(\mathcal{K},[\cdot,\cdot])$. Now (i) follows immediately since $(\tilde{A}_\mathfrak{a}x,y) = [J\tilde{A}_\mathfrak{a}x,y] = [A_\mathfrak{a}x,y]$ for $x \in \mathscr{D}(A_\mathfrak{a}) = \mathscr{D}(\tilde{A}_\mathfrak{a})$ and $y \in \mathscr{D}(\mathfrak{a})$. Statement (ii) is obvious because $\mathscr{D}(A_\mathfrak{a}) = \mathscr{D}(\tilde{A}_\mathfrak{a})$, and so is statement (iii); for $z' \in \mathcal{K}$ we use the identity $(z',y) = [Jz',y] = [z,y]$, with $z := Jz'$, and hence $A_\mathfrak{a}x = J\tilde{A}_\mathfrak{a}x = Jz' = z$ for $x \in \mathscr{D}(A_\mathfrak{a})$. ∎

The proofs of the following results are similar to the Hilbert space case, compare [Kat95, Paragraph VI.2].

Corollary 2.10. *If a form $\mathfrak{a}_0$ is induced by the operator $A_\mathfrak{a}$ in the Krein space in Theorem 2.9 by $\mathfrak{a}_0[x,y] = [A_\mathfrak{a}x,y]$, $x,y \in \mathscr{D}(\mathfrak{a}_0) = \mathscr{D}(A_\mathfrak{a})$, then the form $\mathfrak{a}$ in Theorem 2.9 is the closure of $\mathfrak{a}_0$.*

Proof. Corollary 2.10 is immediate from statement (ii) of Theorem 2.9. ∎

Corollary 2.11. *If B is a linear operator such that $\mathscr{D}(B) \subset \mathscr{D}(\mathfrak{a})$ and $\mathfrak{a}[x,y] = [Bx,y]$ for every $x \in \mathscr{D}(B)$ and every y belonging to a core of $\mathfrak{a}$, then $B \subset A_\mathfrak{a}$.*

Proof. Corollary 2.11 is a direct consequence of statement (iii) of Theorem 2.9. If $x \in \mathscr{D}(B) \subset \mathscr{D}(\mathfrak{a})$, then we have $\mathfrak{a}[x,y] = [z,y]$ with $z := Bx$. By (iii) of Theorem 2.9, $x \in \mathscr{D}(A_\mathfrak{a})$ and $A_\mathfrak{a}x = z = Bx$, i.e., $B \subset A_\mathfrak{a}$. ∎

Proposition 2.12. *The mapping $\mathfrak{a} \mapsto A = A_\mathfrak{a}$ is a one-to-one correspondence between the set of all densely defined, closed sectorial forms and the set of all m-sectorial operators in the Krein space $(\mathcal{K},[\cdot,\cdot])$. The form $\mathfrak{a}$ is bounded if and only if A is bounded, and $\mathfrak{a}$ is symmetric if and only if A is self-adjoint in the Krein space $(\mathcal{K},[\cdot,\cdot])$.*

Proof. First we show that the mapping $\mathfrak{a} \mapsto A = A_\mathfrak{a}$ is injective. Let $\mathfrak{a}_1$ and $\mathfrak{a}_2$ be densely defined, closed and sectorial sesquilinear forms. Then, by (i) of Theorem 2.9, $A_{\mathfrak{a}_1} = A_{\mathfrak{a}_2}$ implies that $\mathfrak{a}_1 = \mathfrak{a}_2$ restricted to $\mathscr{D}(A_{\mathfrak{a}_1}) = \mathscr{D}(A_{\mathfrak{a}_2})$. If we define the forms $\mathfrak{a}_{1_0}[x,y] = [A_{\mathfrak{a}_1}x,y]$ with $\mathscr{D}(\mathfrak{a}_{1_0}) = \mathscr{D}(A_{\mathfrak{a}_1})$ and $\mathfrak{a}_{2_0}[x,y] = [A_{\mathfrak{a}_2}x,y]$ with $\mathscr{D}(\mathfrak{a}_{2_0}) = \mathscr{D}(A_{\mathfrak{a}_2})$, then

$$\mathfrak{a}_1 = \overline{\mathfrak{a}_{1_0}} = \overline{\mathfrak{a}_{2_0}} = \mathfrak{a}_2.$$

It remains to show that the mapping $\mathfrak{a} \mapsto A = A_\mathfrak{a}$ is surjective. Let A be an m-sectorial operator in $\mathscr{K}$. As we have seen in Corollary 2.10, we obtain a densely defined, closed sectorial form $\mathfrak{a}$ as the closure of the densely defined sectorial form

$$\mathfrak{a}_0 [\![x,y]\!] = [Ax,y], \quad \mathscr{D}(\mathfrak{a}_0) = \mathscr{D}(A).$$

Denote the m-sectorial operator of Theorem 2.9 by $A_\mathfrak{a}$. Then, by Corollary 2.11, $A_\mathfrak{a} \supset A$. Hence, since $A_\mathfrak{a}$ and A are m-sectorial in $(\mathscr{K},[\cdot,\cdot])$, we obtain $A_\mathfrak{a} = A$. ∎

Remark 2.13. The uniquely defined operator $A_\mathfrak{a}$ in Theorem 2.9 is called the ***m-sectorial operator*** or simply ***the operator associated with $\mathfrak{a}$ in the Krein space*** $(\mathscr{K},[\cdot,\cdot])$. Vice versa, the form $\mathfrak{a}$ is called ***the form associated with A***.

Next we present the Krein space version of the second representation theorem which can be found in [Kat95], see Theorem VI.2.23 therein.

Theorem 2.14 (The second representation theorem). *Let $\mathfrak{a}$ be a densely defined, closed and non-negative sesquilinear form in a Krein space $(\mathscr{K},[\cdot,\cdot])$. Let $A = A_\mathfrak{a}$ be the associated self-adjoint operator in $(\mathscr{K},[\cdot,\cdot])$. Then we have $\mathscr{D}(\mathfrak{a}) = \mathscr{D}\big((JA)^{1/2}\big)$ and*

$$\mathfrak{a}[\![x,y]\!] = \big((JA)^{1/2}x,(JA)^{1/2}y\big), \quad x,y \in \mathscr{D}(\mathfrak{a}).$$

Furthermore, a subspace $\mathscr{D}'$ of $\mathscr{D}(\mathfrak{a})$ is a core of $\mathfrak{a}$ if and only if it is a core of $(JA)^{1/2}$.

Proof. The proof follows immediately from the Hilbert space version of the theorem, compare [Kat95, Theorem VI.2.23] if we note that the operator associated with $\mathfrak{a}$ in the Hilbert space $(\mathscr{K},(\cdot,\cdot))$ is JA, compare (2.1). ∎

Remark 2.15. The unique m-accretive square root $(JA)^{1/2}$ in the Hilbert space $(\mathscr{K},(\cdot,\cdot))$ of JA is self-adjoint and non-negative in $(\mathscr{K},(\cdot,\cdot))$ such that $\big((JA)^{1/2}\big)^2 = (JA)$ and $\mathscr{D}(JA)$ $(= \mathscr{D}(A))$ is a core of $(JA)^{1/2}$, see [Kat95, Paragraph V.3.11].

Definition 2.16. Let A be a densely defined, sectorial operator in a Krein space $(\mathcal{K}, [\cdot, \cdot])$ and let $\mathfrak{a}_0$ be the closable form associated with A by

$$\mathfrak{a}_0[x, y] := [Ax, y], \quad x, y \in \mathscr{D}(\mathfrak{a}_0) = \mathscr{D}(A).$$

Then the operator $A_F := A_{\overline{\mathfrak{a}_0}} \supset A$ is called **Friedrichs extension** of A in the Krein space $(\mathcal{K}, [\cdot, \cdot])$.

Remark 2.17. If A is a symmetric operator in a Krein space $(\mathcal{K}, [\cdot, \cdot])$ such that A is bounded from below in $(\mathcal{K}, [\cdot, \cdot])$, then A is sectorial and hence the Friedrichs extension A_F of A exists. By [Kat95, Theorem VI.2.6], JA_F is self-adjoint in the Hilbert space $(\mathcal{K}, (\cdot, \cdot))$, that is, A_F is self-adjoint in the Krein space $(\mathcal{K}, [\cdot, \cdot])$.

Remark 2.18. Since an m-sectorial operator in a Krein space has no proper sectorial extension (see Definition 2.1), the Friedrichs extension of an m-sectorial operator A in a Krein space is A itself.

Definition 2.19. Let $\mathfrak{a}_0$ and $\mathfrak{v}$ be closed and sectorial forms in a Krein space $(\mathcal{K}, [\cdot, \cdot])$. By [Kat95, Theorem VI.1.31], the same is true for their sum $\mathfrak{a}_1 = \mathfrak{a}_0 + \mathfrak{v}$. Thus if $\mathfrak{a}_1$ is densely defined, then the m-sectorial operators A_1, A_0 and V associated with $\mathfrak{a}_1$, $\mathfrak{a}_0$ and $\mathfrak{v}$, respectively, are defined. A_1 is the "sum" of A_0 and V in a generalized sense, which is called the **form sum** of A_0 and V and indicated by writing $A_1 = A_0 \dotplus V$.

Remark 2.20.   (i) If any two operators A_0 and V are self-adjoint and bounded from below in a Krein space $(\mathcal{K}, [\cdot, \cdot])$, then the associated forms $\mathfrak{a}_0$ and $\mathfrak{v}$ exist. In this case the form sum A_1 of A_0 and V is the operator associated with the form $\mathfrak{a}_1 = \mathfrak{a}_0 + \mathfrak{v}$ provided $\mathfrak{a}_1$ is densely defined. The requirement of $\mathfrak{a}_1$ being densely defined is weaker than the condition for the ordinary operator sum $S := A_0 + V$ to be densely defined. In some cases it may not be possible to define the operator sum, e.g., $\mathscr{D}(A_1) \cap \mathscr{D}(A_0) = \{0\}$ may occur, while the form sum is defined (an example can be found in [Wei00, Übung 4.17]). Even when S is densely defined it may not be self-adjoint in $(\mathcal{K}, [\cdot, \cdot])$ or have a self-adjoint closure. From (iii) of Theorem 2.9 it follows that $A_1 = A_0 \dotplus V$ is an extension of the operator sum $A_0 + V$.

(ii) If A_0 and V are self-adjoint and bounded from below in $(\mathcal{K}, [\cdot, \cdot])$ such that $A_0 + V$ is densely defined, the Friedrichs extension A_F of $A_0 + V$ exists. In general, however, it is not true that $A_F = A_0 \dotplus V$, see [Kat95, Example VI.2.19].

For the preceding remark compare also [Kat95, Section VI.2.5] and [EE87, Section IV.2].

2.1.3 Relatively Form-Bounded and Relatively Form-Compact Operators

The first representation theorem, Theorem 2.9, gives a one-to-one correspondence between densely defined, closed sectorial forms and m-sectorial operators in a Krein space $(\mathcal{K}, [\cdot, \cdot])$. Using this correspondence, we extend the notion of relative boundedness for forms (see [Kat95, Section VI.1.6 ff.] and Definition 2.21 below) to the associated m-sectorial operators in a Krein space $(\mathcal{K}, [\cdot, \cdot])$. Definitions of relatively form-bounded operators in Hilbert spaces may be found in [Tes09, Section 6.5], [RS75, Section X.2] and [BH04].

The following definition was given in [Kat95, Section VI.1.6].

Definition 2.21. Let $\mathfrak{a}$ and $\mathfrak{v}$ be sesquilinear forms in a Krein space $(\mathcal{K}, [\cdot, \cdot])$. Suppose that $\mathfrak{a}$ is sectorial. If $\mathscr{D}(\mathfrak{a}) \subset \mathscr{D}(\mathfrak{v})$ and there exist non-negative constants α and β such that

$$(2.2) \qquad |\mathfrak{v}[\![x]\!]| \leq \alpha \|x\|^2 + \beta |\mathfrak{a}[\![x]\!]|, \quad x \in \mathscr{D}(\mathfrak{a}),$$

then $\mathfrak{v}$ is called **relatively bounded with respect to** $\mathfrak{a}$. The greatest lower bound β_0 of all possible constants β in (2.2) is called $\mathfrak{a}$-**bound** of $\mathfrak{v}$.

Remark 2.22. If in Definition 2.21 both $\mathfrak{a}$ and $\mathfrak{v}$ are sectorial and closable, then, according to [Kat95, Paragraph VI.1.6], inequality (2.2) also holds for their closures with the same constants α and β.

Definition 2.23. Let A be a self-adjoint and V a symmetric operator in a Krein space $(\mathcal{K}, [\cdot, \cdot])$ such that A and V are bounded from below in $(\mathcal{K}, [\cdot, \cdot])$. Let $\mathfrak{a}$ and $\mathfrak{v}$ be the forms associated with A and with the Friedrichs extension V_F of V, respectively. If $\mathscr{D}(\mathfrak{a}) \subset \mathscr{D}(\mathfrak{v})$ and there exist non-negative constants α and β such that

$$(2.3) \qquad |\mathfrak{v}[\![x]\!]| \leq \alpha \|x\|^2 + \beta |\mathfrak{a}[\![x]\!]|, \quad x \in \mathscr{D}(\mathfrak{a}),$$

then V is called ***relatively form-bounded with respect to A in the Krein space*** $(\mathscr{K},[\cdot,\cdot])$. The greatest lower bound β_0 of all possible constants β in (2.3) is called ***relative form-bound of V with respect to A***.

Remark 2.24. If A and V are not only bounded from below but non-negative in the Krein space $(\mathscr{K},[\cdot,\cdot])$, then inequality (2.3) is equivalent to

$$\left\|(JV_{\mathrm{F}})^{1/2}x\right\|^2 \le \alpha\|x\|^2 + \beta\left\|(JA)^{1/2}x\right\|^2, \quad x \in \mathscr{D}\big((JA)^{1/2}\big),$$

with $\mathscr{D}\big((JA)^{1/2}\big) \subset \mathscr{D}\big((JV_{\mathrm{F}})^{1/2}\big)$.

The following correspondence is obvious, compare (2.1):

Remark 2.25. Let A be a self-adjoint and V a symmetric operator in a Krein space $(\mathscr{K},[\cdot,\cdot])$ such that A and V are bounded from below in $(\mathscr{K},[\cdot,\cdot])$ and $\mathscr{D}(\mathfrak{a}) \subset \mathscr{D}(\mathfrak{v})$, where $\mathfrak{a}$ and $\mathfrak{v}$ are the forms associated with A and V_{F}, respectively. Then V is relatively form-bounded with respect to A and relative form-bound β_0 in the Krein space $(\mathscr{K},[\cdot,\cdot])$ if and only if JV is relatively form-bounded with respect to JA and relative form-bound β_0 in the Hilbert space $(\mathscr{K},(\cdot,\cdot))$ in the usual sense.

As in the case of relatively bounded operators, the constant α in equation (2.3) may have to be chosen very large if β is chosen very close to the relative form-bound β_0.

Remark 2.26. Note that if V is relatively form-bounded with respect to A with relative form-bound zero, then for any $\beta > 0$ there exists an $\alpha_\beta > 0$ such that

$$(2.4) \qquad |\mathfrak{v}[x]| \le \alpha_\beta\|x\|^2 + \beta|\mathfrak{a}[x]|, \quad x \in \mathscr{D}(\mathfrak{a}).$$

The following theorem shows that relative form-boundedness is weaker than relative boundedness, see [Kat95, Theorem VI.1.38].

Theorem 2.27. *Let A be a self-adjoint and V a symmetric operator in a Krein space $(\mathscr{K},[\cdot,\cdot])$ such that A is bounded from below in $(\mathscr{K},[\cdot,\cdot])$. Suppose that V is relatively bounded with respect to A and A-bound β_0. Then V is relatively form-bounded with respect to A with relative form-bound $\le \beta_0$. In particular, if V has A-bound 0, then also the relative form-bound of V with respect to A is 0.*

Proof. According to Remark 1.17 we have

$$\|JVx\| \le \alpha\|x\| + \beta\|JAx\|, \quad x \in \mathscr{D}(A),$$

where $\alpha \ge 0$ and $\beta \ge \beta_0 \ge 0$. A is bounded from below in $(\mathscr{K},[\cdot,\cdot])$ and thus JA is bounded from below in $(\mathscr{K},(\cdot,\cdot))$. Hence, by [Kat95, Theorem V.4.11], the operator JA_ε, where $A_\varepsilon := A + \varepsilon V$, $\varepsilon \in \mathbb{R}$, is self-adjoint and bounded from below in the Hilbert space $(\mathscr{K},(\cdot,\cdot))$ if $|\varepsilon| < \beta^{-1}$ and the lower bound γ_{JA_ε} of JA_ε satisfies

$$\gamma_{JA_\varepsilon} \ge \gamma_{JA} - |\varepsilon| \max\left\{ \frac{\alpha}{1-\beta|\varepsilon|}, \alpha + \beta|\gamma_{JA}| \right\},$$

where γ_{JA} is the lower bound of JA. Hence

$$(2.5) \quad -\varepsilon[Vx,x] = -\varepsilon(JVx,x) \le -\gamma_{JA_\varepsilon}\|x\|^2 + (JAx,x) = -\gamma_{JA_\varepsilon}\|x\|^2 + [Ax,x],$$

for $x \in \mathscr{D}(A)$.

Case (i): $[Vx,x] \ge 0$. Since ε can be chosen arbitrarily close to $-\beta^{-1}$ and thus to $-\beta_0^{-1}$, we obtain from (2.5) with $\varepsilon_0 := -\varepsilon > 0$

$$[Vx,x] = \big|[Vx,x]\big| \le \max\left\{ 0, \frac{-\gamma_{JA_\varepsilon}}{\varepsilon_0} \right\}\|x\|^2 + \frac{1}{\varepsilon_0}\big|[Ax,x]\big|, \quad x \in \mathscr{D}(A).$$

Case (ii): $[Vx,x] < 0$. Since ε can be chosen arbitrarily close to β^{-1} and thus to β_0^{-1}, we obtain from (2.5) with $\varepsilon_0 := \varepsilon > 0$

$$-[Vx,x] = \big|[Vx,x]\big| \le \max\left\{ 0, \frac{-\gamma_{JA_\varepsilon}}{\varepsilon_0} \right\}\|x\|^2 + \frac{1}{\varepsilon_0}\big|[Ax,x]\big|, \quad x \in \mathscr{D}(A).$$

Altogether, the form $[Vx,x]$ is relatively bounded with respect to the form $[Ax,x]$ with relative form-bound $\le \beta_0$. By Remark 2.22, the same is true for their closures, that is, V is relatively form-bounded with respect to A with relative form-bound less than or equal to β_0. ∎

Definition and Remark 2.28. Let A be a self-adjoint operator in a Krein space $(\mathscr{K},[\cdot,\cdot])$ such that A is non-negative in $(\mathscr{K},[\cdot,\cdot])$. Set

$$(x,y)_\mathfrak{a} := (x,y) + \mathfrak{a}[x,y], \quad x,y \in \mathscr{D}(\mathfrak{a}),$$

where $\mathfrak{a}$ is the form associated with A. Then $(\cdot,\cdot)_\mathfrak{a}$ defines a positive definite inner product on $\mathscr{D}(\mathfrak{a})$ and $(\mathscr{D}(\mathfrak{a}),(\cdot,\cdot)_\mathfrak{a})$ becomes a Hilbert space, which we denote by $\mathscr{D}_\mathfrak{a}$ (compare Definition and Remark 1.14). The norm which is induced by $(\cdot,\cdot)_\mathfrak{a}$ is denoted by $\|\cdot\|_\mathfrak{a}$ and $(\cdot,\cdot)_\mathfrak{a}$ is called **graph inner product**.

Remark 2.29. Let A be a self-adjoint and V a symmetric operator in a Krein space $(\mathcal{K},[\cdot,\cdot])$ such that A and V are non-negative in $(\mathcal{K},[\cdot,\cdot])$ and $\mathcal{D}(\mathfrak{a}) \subset \mathcal{D}(\mathfrak{v})$, where $\mathfrak{a}$ and $\mathfrak{v}$ are the forms associated with A and V_F, respectively. Suppose that V is relatively form-bounded with respect to A. Then the restriction $\widehat{(JV)}^{1/2} := (JV)^{1/2}\big|_{\mathcal{D}(\mathfrak{a})}$ as an operator from $\mathcal{D}_\mathfrak{a}$ to $\mathcal{K}$ is bounded.

Proof. By assumption, inequality (2.3) holds. Define $\mu := \max\{\alpha,\beta\}$. Then we have

$$(2.6) \quad \left\|\widehat{(JV)}^{1/2}x\right\|^2 = \left\|(JV)^{1/2}x\right\|^2 = |\mathfrak{v}[\![x]\!]| \le \mu\|x\|^2 + \mu|\mathfrak{a}[\![x]\!]| = \mu\|x\|_\mathfrak{a}^2, \quad x \in \mathcal{D}(\mathfrak{a}),$$

i.e., $\widehat{(JV)}^{1/2}$ is bounded.

Vice versa, if $\widehat{(JV)}^{1/2}$ is bounded from $\mathcal{D}_\mathfrak{a}$ to $\mathcal{K}$, then

$$\left\|\widehat{(JV)}^{1/2}x\right\|^2 \le \left\|\widehat{(JV)}^{1/2}\right\|^2 \|x\|_\mathfrak{a}^2 = \alpha\|x\|^2 + \beta\mathfrak{a}[\![x]\!]|, \quad x \in \mathcal{D}(\mathfrak{a}),$$

where $\alpha := \beta := \left\|\widehat{(JV)}^{1/2}\right\|^2$. ∎

As in Chapter 1, we are particularly interested in relatively form-bounded operators with relative form-bound 0. A sufficient condition is given by Lemma 2.31 below. In the following definition of relatively form-compact operators we assume for simplicity that A and V are non-negative in a Krein space; a more general definition can be found, e.g., in [GMMN09, Definition 3.1].

Definition 2.30. Let A be a self-adjoint and V a symmetric operator in a Krein space $(\mathcal{K},[\cdot,\cdot])$ such that A and V are non-negative in $(\mathcal{K},[\cdot,\cdot])$. Let $\mathfrak{a}$ and $\mathfrak{v}$ be the forms associated with A and V_F, respectively. Let V be relatively form-bounded with respect to A in the Krein space $(\mathcal{K},[\cdot,\cdot])$. Then V is called *relatively form-compact with respect to A* if and only if the restriction $(JV)^{1/2}\big|_{\mathcal{D}(\mathfrak{a})}$ as an operator from $\mathcal{D}_\mathfrak{a}$ to $\mathcal{K}$ is a compact map.

The following theorem is the analogue of Theorem 1.22 for relatively form-compact operators.

Lemma 2.31. *Let A be a self-adjoint and V a symmetric operator in a Krein space $(\mathcal{K},[\cdot,\cdot])$ such that A and V are non-negative in $(\mathcal{K},[\cdot,\cdot])$. If V is relatively form-compact with respect to A, then the relative form-bound of V with respect to A is 0.*

Proof. The proof is similar to the proof of Theorem 1.22. We prove the theorem by contradiction. Suppose that the relative form-bound of V is not 0. Then there exist an $\varepsilon > 0$ and a sequence $(x_n)_{n=1}^{\infty} \subset \mathscr{D}\big((JA)^{1/2}\big)$ such that for all natural numbers n

$$(2.7) \qquad \big\|(JV)^{1/2}x_n\big\|^2 > n\|x_n\|^2 + \varepsilon\big\|(JA)^{1/2}x_n\big\|^2.$$

Set $y_n := x_n/\|x_n\|_a$. Then, by inequalities (2.3), (2.6) and (2.7), and since $\|y_n\|_a = 1$, $n \in \mathbb{N}$, we have for $n \geq \varepsilon$

$$(2.8) \qquad \begin{aligned} \max\{\alpha,\beta\} = \max\{\alpha,\beta\}\|y_n\|_a^2 &\geq \big\|(JV)^{1/2}y_n\big\|^2 \\ &> n\|y_n\|^2 + \varepsilon\big\|(JA)^{1/2}y_n\big\|^2 \\ &\geq \varepsilon\Big(\|y_n\|^2 + \big\|(JA)^{1/2}y_n\big\|^2\Big), \end{aligned}$$

where α and β are constants according to (2.3). Hence, by inequality (2.8), $y_n \to 0$ if $n \to \infty$ and $(y_n)_{n=1}^{\infty}$ and $\big((JA)^{1/2}y_n\big)_{n=1}^{\infty}$ are bounded sequences. Thus, since V is relatively form-compact with respect to A, there exists a subsequence $(y_{n_k})_{k=1}^{\infty}$ of $(y_n)_{n=1}^{\infty}$ such that $\big(V^{1/2}y_{n_k}\big)_{k=1}^{\infty}$ converges to some $z \in \mathscr{K}$. Since V be closable, $(y_{n_k})_{k=1}^{\infty}$ converges to zero and $\big(V^{1/2}y_{n_k}\big)_{k=1}^{\infty}$ converges to z, it follows that $z = 0$. By inequality (2.8), also $\big(A^{1/2}y_{n_k}\big)_{k=1}^{\infty}$ converges to 0. This leads to the contradiction

$$1 = \|y_{n_k}\|_a^2 = \|y_{n_k}\|^2 + \|A^{1/2}y_{n_k}\|^2 \longrightarrow 0, \quad k \to 0. \qquad \blacksquare$$

2.2 Continuity of Separated Parts of the Spectrum

If V is relatively form-bounded with respect to A_0 with A_0-form-bound less than 1, then the spectrum of $A_0 + \varepsilon V$, $\varepsilon \in [0,1]$, cannot suddenly expand for increasing ε. As in Chapter 1, we are interested in the case when the spectrum $\sigma(A_0)$ consists only of isolated eigenvalues or isolated parts. In the first paragraph we present some well-known perturbation results for self-adjoint operators in a Hilbert space. In the second paragraph of this section we extend the obtained results to the case of self-adjoint operators in Krein spaces. Analogously to Chapter 1, if the spectrum $\sigma(A_0)$ is real, then our main results state conditions which guarantee that also the spectrum of $A_0 + V$ is real, even when $A_0 + V$ is not self-adjoint in a Hilbert space.

2.2.1 Perturbation of Spectra of Self-Adjoint Operators in Hilbert Spaces

In this paragraph we consider the case where the spectrum $\sigma(A_0)$ of a self-adjoint operator A_0 in a Hilbert space $\mathcal{H}$ contains a bounded subset $\sigma_1(A_0)$ separated from the rest $\sigma(A_0)\backslash\sigma_1(A_0)$ by a rectifiable closed curve Γ. For symmetric perturbations V which are relatively form-bounded with respect to A_0, we state conditions for the spectrum $A_0 + V$ to be also separated into two parts by Γ.

We recall the following theorem, which was proved in [Kat95, Theorem VI.3.9].

Theorem 2.32 (Cf. [Kat95, Theorem VI.3.9 and Remark VI.3.10]). *In a Hilbert space $\mathcal{H}$ let $\mathfrak{a}_0$ be a densely defined, closed symmetric form bounded from below with associated self-adjoint operator A_0. Let $\mathfrak{v}$ be a form relatively bounded with respect to $\mathfrak{a}_0$ so that $\mathscr{D}(\mathfrak{v}) \supset \mathscr{D}(\mathfrak{a}_0)$ and*

$$|\mathfrak{v}[x]| \leq \alpha\|x\|^2 + \beta\mathfrak{a}_0[x], \quad x \in \mathscr{D}(\mathfrak{a}_0),$$

where $0 \leq \beta < 1$ but α may be positive, negative or zero. Then $\mathfrak{a}_1 = \mathfrak{a}_0 + \mathfrak{v}$ is sectorial and closed. The associated operator A_1 is m-sectorial; A_1 is self-adjoint if $\mathfrak{v}$ is symmetric. Let $\kappa = 1$ if $\mathfrak{v}$ symmetric, otherwise $\kappa = 2$. If $z \in \rho(A_0)$ and

$$(2.9) \qquad \kappa\left\|(\alpha + \beta A_0)(A_0 - z)^{-1}\right\| < 1,$$

then $z \in \rho(A_1)$ and

$$(2.10) \qquad \left\|(A_1 - z)^{-1} - (A_0 - z)^{-1}\right\| < \frac{4\kappa\left\|(\alpha + \beta A_0)(A_0 - z)^{-1}\right\|\left\|(A_0 - z)^{-1}\right\|}{\left(1 - \kappa\left\|(\alpha + \beta A_0)(A_0 - z)^{-1}\right\|\right)^2}.$$

For symmetric $\mathfrak{v}$, results similar to Theorem 2.32, which also show that A_1 is self-adjoint, can be found in [Sim71b, Theorem II.7] (KLMN[1] theorem) or [Nel64, Appendix], but they do not include inequality (2.10).

In the following we apply Theorem 2.32 to families of operators $A_\varepsilon = A_0 + \varepsilon V$, $\varepsilon \in [0, 1]$, to establish convergence in the generalized sense for $\varepsilon \to 0$.

---

[1]KLMN stands for Kato, Lions, Lax, Milgram and Nelson, compare [Sim71a].

Remark 2.33. If in Theorem 2.32, instead of $\mathfrak{a}_1$, the sum of forms $\mathfrak{a}_\varepsilon :=$ $\mathfrak{a}_0 + \varepsilon \mathfrak{v}$, $\varepsilon \in [0,1]$, is considered, then inequality (2.10) becomes (see the proof of [Kat95, Theorem VI.3.9] or Theorem 2.39 below)

$$\left\| (A_\varepsilon - z)^{-1} - (A_0 - z)^{-1} \right\| < \varepsilon \frac{4\kappa \left\| (\alpha + \beta A_0)(A_0 - z)^{-1} \right\| \left\| (A_0 - z)^{-1} \right\|}{\left(1 - \kappa \left\| (\alpha + \beta A_0)(A_0 - z)^{-1} \right\| \right)^2}.$$

Hence

$$\left\| (A_\varepsilon - z)^{-1} - (A_0 - z)^{-1} \right\| \longrightarrow 0, \quad \varepsilon \to 0,$$

and, by Theorem 1.27, A_ε converges to A_0 in the generalized sense.

Corollary 2.34. *Let A_0 be a self-adjoint and V a symmetric operator in a Hilbert space $\mathscr{H}$ such that A_0 and V are bounded from below in $\mathscr{H}$. Further, let V be relatively form-bounded with respect to A_0 such that (2.3) holds with constants $\alpha \in \mathbb{R}$ and $\beta \in [0,1)$. Define the self-adjoint family of operators A_ε, $\varepsilon \in [0,1]$, as the operator associated with the sum of forms $\mathfrak{a}_\varepsilon := \mathfrak{a}_0 + \varepsilon \mathfrak{v}$, $\varepsilon \in [0,1]$, with $\mathfrak{a}_0$ and $\mathfrak{v}$ being the forms associated with A_0 and V_F, respectively. Suppose that the spectrum of A_0 is separated into two parts $\sigma_1(A_0)$ and $\sigma_2(A_0)$ by a Cauchy contour Γ. If*

$$(2.11) \qquad\qquad \sup_{z \in \Gamma} \left\| (\alpha + \beta A_0)(A_0 - z)^{-1} \right\| < 1,$$

then the spectrum of $A_1 = A_0 \dotplus V$ is likewise separated into two parts $\sigma_1(A_1)$ and $\sigma_2(A_1)$, $\Gamma \subset \rho(A_1)$ and the results of Theorem 1.35 hold.

Proof. The assumptions of Theorem 2.32 are satisfied. Hence $\Gamma \subset \rho(A_1)$. Define the isolated part $\sigma_1(A_1) := \operatorname{int}\Gamma \cap \sigma(A_1)$ of $\sigma(A_1)$ and $\sigma_2(A_1) := \sigma(A_1) \backslash \sigma_1(A_1)$. By Lemma 1.33, A_1 can be decomposed according to $\mathscr{H} = \mathscr{M}_1(A_1) \oplus \mathscr{M}_2(A_1)$. We have $\mathscr{M}_1 = E(A_1, \sigma_1(A_1))\mathscr{H}$ and $\mathscr{M}_2 = \big(1 - E(A_1, \sigma_1(A_1))\big)\mathscr{H}$. According to Remark 2.33, $(A_\varepsilon - z)^{-1}$, $z \in \Gamma$, depends continuously on ε for $0 \leq \varepsilon \leq 1$. Hence the Riesz projection $E(A_\varepsilon, \sigma_1(A_\varepsilon))$ of A_ε corresponding to $\sigma_1(A_\varepsilon)$ is continuous in ε for $0 \leq \varepsilon \leq 1$. The last part of the proof is analogous to that of Theorem 1.35. ∎

Theorem 2.32 and Corollary 2.34 provide the necessary tools to obtain results analogous to those for relatively bounded operators in the first chapter, see Theorems 1.39 and 1.40. Again we distinguish the following situations:

(a) We consider one isolated eigenvalue (an infinite sequence of isolated eigenvalues, respectively) of the unperturbed operator A_0.

(b) We consider one isolated compact part (an infinite sequence of isolated compact parts, respectively) of the spectrum of the unperturbed operator A_0.

Note that in the theorems below, in contrast to Theorems 1.39 and 1.40, the operator A_1 is self-adjoint in the Hilbert space by Theorem 2.32 since V is symmetric in the Hilbert space.

In situation (a) we obtain the following result.

Theorem 2.35. *Let A_0 be a self-adjoint and V a symmetric operator in a Hilbert space $\mathscr{H}$ such that A_0 and V are bounded from below in $\mathscr{H}$. Further, let V be relatively form-bounded with respect to A_0 with relative form-bound less than $1/2$ and let A_ε, $\varepsilon \in [0, 1]$, be the family of self-adjoint operators in $\mathscr{H}$ which is associated to the sum of forms $\mathfrak{a}_\varepsilon := \mathfrak{a}_0 + \varepsilon \mathfrak{v}$, $\varepsilon \in [0, 1]$, with $\mathfrak{a}_0$ and $\mathfrak{v}$ being the forms associated to A_0 and V_F, respectively.*

(i) *Let $\lambda^0 \in \mathbb{R}$ be an isolated eigenvalue of A_0 with multiplicity $m < \infty$ and set $\delta := \operatorname{dist}(\lambda^0, \sigma(A_0) \setminus \{\lambda^0\})$. If (2.3) holds with constants $\alpha \geq 0$ and $\beta \in [0, 1/2)$ such that*

$$(2.12) \qquad \frac{1}{\delta}\left(\alpha + \beta\left(\delta + |\lambda^0|\right)\right) < \frac{1}{2},$$

then $\sigma_1 := \sigma(A_1) \cap \left(\lambda^0 - \delta/2, \lambda^0 + \delta/2\right)$ consists of a finite system of isolated (real) eigenvalues with total multiplicity m.

(ii) *Let A_0 have discrete spectrum consisting of eigenvalues $\lambda_1^0 < \lambda_2^0 < \cdots$ with multiplicities $m_n < \infty$, $n \in \mathbb{N}$ and set $\delta_n := \operatorname{dist}(\lambda_n^0, \sigma(A_0) \setminus \{\lambda_n^0\})$, $n \in \mathbb{N}$. If (2.3) holds with constants $\alpha_n \geq 0$ and $\beta_n \in [0, 1/2)$, $n \in \mathbb{N}$, such that*

$$(2.13) \qquad \gamma := \sup_{n \in \mathbb{N}}\left(\frac{1}{\delta_n}\left(\alpha_n + \beta_n\left(\delta_n + |\lambda_n^0|\right)\right)\right) < \infty,$$

then $\sigma_{n,\varepsilon} := \sigma(A_\varepsilon) \cap \left(\lambda_n^0 - \delta_n/2, \lambda_n^0 + \delta_n/2\right)$ consists of a finite system of isolated (real) eigenvalues with total multiplicity m_n for all $n \in \mathbb{N}$ and $\varepsilon \in [0, \varepsilon_0]$, where $\varepsilon_0 \in (0, 1]$ has to be chosen such that $\varepsilon_0 < 1/(2\gamma)$.

Proof. (i). As in the proof of Theorem 1.39, let Γ be the positively oriented curve along the circle with center λ^0 and radius $\delta/2$. Then $\Gamma \subset \rho(A_0)$, $\{\lambda^0\} \subset \operatorname{int}(\Gamma)$ and $\left(\Gamma \cup \operatorname{int}(\Gamma)\right) \cap \left(\sigma(A_0) \setminus \{\lambda^0\}\right) = \emptyset$. According to the proof of Theorem 1.39, inequality (2.11) is satisfied if

$$\alpha \frac{2}{\delta} + \beta\left(2 + \frac{2|\lambda^0|}{\delta}\right) < 1,$$

or, equivalently, (2.12) holds. By Corollary 2.34, the spectrum of A_ε is separated into the two parts $\sigma_\varepsilon := \operatorname{int}\Gamma \cap \sigma(A_\varepsilon)$ and $\sigma(A_\varepsilon)\setminus\sigma_\varepsilon$ such that σ_ε consists of isolated eigenvalues with total multiplicity m for all $\varepsilon \in [0,1]$.

(ii). By (2.13), the assumptions of (i) are satisfied for A_0 and εV with $\varepsilon \in [0,\varepsilon_0]$ for every eigenvalue λ_n^0, $n \in \mathbb{N}$, if we choose ε_0 such that

$$\varepsilon_0 \sup_{n\in\mathbb{N}}\left(\frac{1}{\delta_n}\left(\alpha_n + \beta_n\big(\delta_n + |\lambda_n^0|\big)\right)\right) < \frac{1}{2}. \qquad \blacksquare$$

The following theorem deals with situation (b), where isolated compact parts of the spectrum of the unperturbed operator are considered.

Theorem 2.36. *Let A_0 be a self-adjoint and V a symmetric operator in a Hilbert space $\mathcal{H}$ such that A_0 and V are bounded from below in $\mathcal{H}$. Further, let V be relatively form-bounded with respect to A_0 with relative form-bound less than $1/2$ and let A_ε, $\varepsilon \in [0,1]$, be the family of self-adjoint operators in $\mathcal{H}$ which is associated to the sum of forms $\mathfrak{a}_\varepsilon := \mathfrak{a}_0 + \varepsilon\mathfrak{v}$, $\varepsilon \in [0,1]$, with $\mathfrak{a}_0$ and $\mathfrak{v}$ being the forms associated to A_0 and V_F, respectively.*

(i) *Let σ_0 be an isolated part of $\sigma(A_0)$ such that $\sigma_0 = \sigma(A_0)\cap[\lambda^-,\lambda^+]$ with $\lambda^-,\lambda^+ \in \mathbb{R}$ and set $\delta := \operatorname{dist}(\sigma_0,\sigma(A_0)\setminus\sigma_0)$. If (2.3) holds with constants $\alpha \geq 0$ and $\beta \in [0,1/2)$ such that*

$$(2.14) \qquad \frac{1}{\delta}\left(\alpha + \beta\big(\delta + \max\{|\lambda^-|,|\lambda^+|\}\big)\right) < \frac{1}{2},$$

then $\sigma_\varepsilon := \sigma(A_\varepsilon)\cap\left(\lambda^- - \delta/2, \lambda^+ + \delta/2\right)$ is an isolated part of $\sigma(A_\varepsilon)$ for all $\varepsilon \in [0,1]$. Furthermore, $\dim E(A_0,\sigma_0)\mathcal{H} = \dim E(A_\varepsilon,\sigma_\varepsilon)\mathcal{H}$ for $\varepsilon \in [0,1]$.

(ii) *Let*

$$(2.15) \qquad \sigma(A_0) = \bigcup_{n\in\mathbb{N}} \sigma_{n,0}$$

with $\sigma_{n,0} = \sigma(A_0)\cap[\lambda_n^-,\lambda_n^+]$, $n \in \mathbb{N}$, where $\lambda_n^-,\lambda_n^+ \in \mathbb{R}$, and

$$\lambda_1^- \leq \lambda_1^+ < \lambda_2^- \leq \lambda_2^+ < \cdots.$$

Define $\delta_n := \operatorname{dist}(\sigma_{n,0},\sigma(A_0)\setminus\sigma_{n,0})$, $n \in \mathbb{N}$. If (2.3) holds with constants $\alpha_n \geq 0$ and $\beta_n \in [0,1/2)$, $n \in \mathbb{N}$, such that

$$\gamma := \sup_{n\in\mathbb{N}}\left(\frac{1}{\delta_n}\left(\alpha_n + \beta_n\big(\delta_n + \max\{|\lambda_n^-|,|\lambda_n^+|\}\big)\right)\right) < \infty,$$

then $\sigma_{n,\varepsilon} := \sigma(A_\varepsilon) \cap \left(\lambda_n^- - \delta_n/2, \lambda_n^+ + \delta_n/2\right)$ *is an isolated part of* $\sigma(A_\varepsilon)$ *for all* $n \in \mathbb{N}$, $\varepsilon \in [0,\varepsilon_0]$, *and*

$$\sigma(A_\varepsilon) = \bigcup_{n \in \mathbb{N}} \sigma_{n,\varepsilon}, \quad \varepsilon \in [0,\varepsilon_0],$$

where $\varepsilon_0 \in (0,1]$ *has to be chosen such that* $\varepsilon_0 < 1/(2\gamma)$. *Furthermore,* $\dim E(A_0, \sigma_{n,0})\mathcal{H} = \dim E(A_\varepsilon, \sigma_{n,\varepsilon})\mathcal{H}$ *for* $\varepsilon \in [0,\varepsilon_0]$ *and* $n \in \mathbb{N}$.

Proof. The proof is analogous to that of Theorem 2.35 and Theorem 1.40. ∎

Remark 2.37. As in Chapter 1, applying similar arguments as in Theorem 2.36 for one spectral gap (a,b) in the spectrum of a self-adjoint operator A_0 in a Hilbert space, we can formulate an analogue statement as in Remark 1.42 for relatively bounded operators. For relatively form-bounded operators in a Hilbert space (and the forms these operators correspond to), condition (1.22) can be relaxed since, by the triangle inequality, condition (2.9) is slightly weaker than the corresponding condition (1.15) for relatively bounded operators.

Theorem 2.38. *Let* A_0 *be a self-adjoint and* V *a symmetric operator in a Hilbert space* $\mathcal{H}$ *such that* A_0 *and* V *are bounded from below in* $\mathcal{H}$. *Let* A_1 *be the operator associated with the sum of forms* $\mathfrak{a}_1 := \mathfrak{a}_0 + \mathfrak{v}$ *with* $\mathfrak{a}_0$ *and* $\mathfrak{v}$ *being the forms associated with* A_0 *and* V_F, *respectively. Suppose the open interval* (a,b) *is a subset of* $\rho(A_0)$. *Let* V *be relatively form-bounded with respect to* A_0 *with relative form-bound less than* 1. *Define the interval* I *by*

$$I := \left(a + \left(\alpha + \beta|a|\right), b - \left(\alpha + \beta|b|\right)\right).$$

If

$$(2.16) \qquad \frac{1}{\delta}\left(\alpha + \beta\frac{|a| + |b|}{2}\right) < \frac{1}{2},$$

where $\alpha \geq 0$ *and* $\beta \in [0,1)$ *are constants according to* (2.3) *and* $\delta = b - a$, *then* $I \neq \emptyset$ *and* I *is a subset of* $\rho(A_1)$.

Proof. $I \neq \emptyset$ if and only if (2.16) is satisfied.

Without loss of generality assume $b > 0$ (otherwise consider the operators $-A_0$ and $-V$). Let $z = x + iy$, such that $x \in (a,b)$ and $y \in \mathbb{R}$. That is, $z \in (a,b) + i\mathbb{R} \subset \rho(A_0)$. Then, according to Theorem 2.32, $z \in \rho(A_1)$ if

$$(2.17) \qquad \left\|(\alpha + \beta A_0)(A_0 - z)^{-1}\right\| < 1.$$

Since A_0 is self-adjoint in the Hilbert space $\mathscr{H}$, it is possible to apply Proposition 1.38. Thus the fact that $|\lambda - x| \leq |\lambda - z|$ for $\lambda \in \sigma(A_0) \subset \big(\mathbb{R} \setminus (a,b)\big)$ implies that condition (2.17) is satisfied if

$$(2.18) \qquad \sup_{\lambda \in \sigma(A_0)} \frac{\alpha + \beta |\lambda|}{|\lambda - x|} < 1.$$

Define the function $f : \mathbb{C} \mapsto \mathbb{R}$ by

$$f(\lambda) := \frac{\alpha + \beta |\lambda|}{|\lambda - x|}.$$

Now the supremum in (2.18) can be estimated (see [Ves08, Theorem 3.1]).

We distinguish the two cases $a > 0$ and $a \leq 0$.

Case (i): $a > 0$. Since $x \in (a,b)$ also $x > 0$.

If $\lambda \leq 0$, then

$$(2.19) \qquad \frac{\mathrm{d}}{\mathrm{d}\lambda} f(\lambda) = \frac{\mathrm{d}}{\mathrm{d}\lambda} \frac{\alpha - \beta\lambda}{x - \lambda} = \frac{-\beta(x - \lambda) + (\alpha - \beta\lambda)}{(x - \lambda)^2} = \frac{\alpha - \beta x}{(x - \lambda)^2}$$

and hence

$$(2.20) \qquad \sup_{\lambda \leq 0} f(\lambda) = \begin{cases} \beta, & \text{if } \alpha \leq \beta x, \\ \dfrac{\alpha}{x}, & \text{if } \beta x < \alpha. \end{cases}$$

If $\lambda \in [0, a]$, then

$$(2.21) \qquad \frac{\mathrm{d}}{\mathrm{d}\lambda} f(\lambda) = \frac{\mathrm{d}}{\mathrm{d}\lambda} \frac{\alpha + \beta\lambda}{x - \lambda} = \frac{\beta(x - \lambda) + (\alpha + \beta\lambda)}{(x - \lambda)^2} = \frac{\alpha + \beta x}{(x - \lambda)^2}$$

and hence

$$(2.22) \qquad \sup_{0 \leq \lambda \leq a} f(\lambda) = \max_{0 \leq \lambda \leq a} f(\lambda) = \frac{\alpha + \beta a}{x - a} > \frac{\alpha}{x}.$$

If $\lambda \geq b$, then

$$(2.23) \qquad \frac{\mathrm{d}}{\mathrm{d}\lambda} f(\lambda) = \frac{\mathrm{d}}{\mathrm{d}\lambda} \frac{\alpha + \beta\lambda}{\lambda - x} = \frac{\beta(\lambda - x) - (\alpha + \beta\lambda)}{(\lambda - x)^2} = \frac{-\alpha - \beta x}{(\lambda - x)^2}$$

and hence

$$(2.24) \qquad \sup_{b \leq \lambda} f(\lambda) = \max_{b \leq \lambda} f(\lambda) = \frac{\alpha + \beta b}{b - x} > \beta.$$

Formulas (2.20), (2.22) and (2.24) now imply

$$\sup_{\lambda \notin (a,b)} f(\lambda) = \max \left\{ \frac{\alpha + \beta a}{x - a}, \frac{\alpha + \beta b}{b - x} \right\}$$

which is less than 1 if $x \in I$.

Case (ii): $a \leq 0$. If $\lambda \leq a$, then, by equation (2.19),

(2.25)
$$\sup_{\lambda \leq a} f(\lambda) = \begin{cases} \beta, & \text{if } \alpha \leq \beta x, \\ \dfrac{\alpha - \beta a}{x - a}, & \text{if } \beta x < \alpha. \end{cases}$$

If $\lambda \geq b$, then, by equation (2.23),

(2.26)
$$\sup_{\lambda \geq b} f(\lambda) = \begin{cases} \beta, & \text{if } \alpha + \beta x \leq 0, \\ \dfrac{\alpha + \beta b}{b - x}, & \text{if } 0 < \alpha + \beta x. \end{cases}$$

In this case equations (2.25) and (2.26) imply

$$\sup_{\lambda \notin (a,b)} f(\lambda) = \max \left\{ \frac{\alpha - \beta a}{x - a}, \frac{\alpha + \beta b}{b - x} \right\}$$

which is less than 1 if $x \in I$. ∎

2.2.2 Perturbation of Spectra of Self-Adjoint Operators in Krein Spaces

In this paragraph we extend the results of the previous section to operators which are self-adjoint in Krein spaces. In the following we generalize Theorem 2.32 proved in [Kat95, Theorem VI.3.9] to the case when the operator A_1 is no longer sectorial. This is needed in order to treat perturbations V which are semi-bounded in a Krein space.

Theorem 2.39. *Let A_0 be a self-adjoint and V a symmetric operator in a Krein space $(\mathcal{K}, [\cdot, \cdot])$ such that A_0 and V are bounded from below in $(\mathcal{K}, [\cdot, \cdot])$ and A_0 is also self-adjoint in the Hilbert space $(\mathcal{K}, (\cdot, \cdot))$. Further, let V be relatively form-bounded with respect to A_0 such that (2.3) holds with constants $\alpha \geq 0$ and $\beta \in [0, 1)$. Then the operator A_1 which corresponds to the sum of* *forms* $\mathfrak{a}_1 := \mathfrak{a}_0 + \mathfrak{v}$, *where* $\mathfrak{a}_0$ *and* $\mathfrak{v}$ *are the forms associated with* A_0 *and* V_F, *respectively, is self-adjoint. If* $z \in \rho(A_0)$ *and*

$$(2.27) \qquad \alpha\|(A_0-z)^{-1}\| + \beta\|A_0(A_0-z)^{-1}\| < 1,$$

then $z \in \rho(A_1)$ *and*

$$(2.28) \quad \|(A_1-z)^{-1}-(A_0-z)^{-1}\| < \frac{4\big(\alpha\|(A_0-z)^{-1}\|+\beta\|A_0(A_0-z)^{-1}\|\big)\|(A_0-z)^{-1}\|}{\Big(1-\big(\alpha\|(A_0-z)^{-1}\|+\beta\|A_0(A_0-z)^{-1}\|\big)\Big)^2}.$$

Proof. Since $A_0 = A_0^{[*]}$ and $V \subset V^{[*]}$ the operators $\widehat{A}_0 := JA_0$ and $\widehat{V} := JV$ are self-adjoint and symmetric, respectively, in the Hilbert space $(\mathscr{K},(\cdot,\cdot))$. Thus the self-adjointness of A_1 follows immediately from Theorem 2.32.

The proof of inequality (2.28) is similar to the proof of [Kat95, Theorem VI.3.9]. Assume that $b > 0$ and set $\mathfrak{a}_0' = \mathfrak{a}_0 + \alpha\beta^{-1} + \delta$ and $\mathfrak{a}_1' = \mathfrak{a}_1 + \alpha\beta^{-1} + \delta$ for some $\delta > 0$ which will be determined later. Here $\mathfrak{a}_0$ and $\mathfrak{a}_1$ are the forms associated with A_0 and A_1 in the Krein space $(\mathscr{K},[\cdot,\cdot])$, respectively. The operators associated with $\mathfrak{a}_0'$ and $\mathfrak{a}_1'$ in the Krein space $(\mathscr{K},[\cdot,\cdot])$ are respectively $A_0' = A_0 + \alpha\beta^{-1}J + \delta J$ and $A_1' = A_1 + \alpha\beta^{-1}J + \delta J$ by equation (2.1). Inequality (2.3) implies $\mathfrak{a}_0 + \alpha\beta^{-1} \geq 0$ and thus $\mathfrak{a}_0' \geq \delta$ in $(\mathscr{K},[\cdot,\cdot])$. By (2.3), we also have $|\mathfrak{v}[x]| \leq \beta\mathfrak{a}_0'[x]$. According to [Kat95, Lemma VI.3.1], $\mathfrak{v}[x,y]$ may be written in the form $(CG'x,G'y)$, $x,y \in \mathscr{D}(G')$, where C is a linear operator with $\|C\| \leq \beta$ and $G' := (JA_0')^{1/2}$. Since $JA_0' \geq \delta$ in $(\mathscr{K},(\cdot,\cdot))$ we have $G' \geq \delta^{1/2}$ in $(\mathscr{K},(\cdot,\cdot))$ and hence $\|G'^{-1}\| \leq \delta^{-1/2}$. Furthermore

$$\mathfrak{a}_0'[x,y] = [A_0'x,y] = (JA_0'x,y) = \big((JA_0')^{1/2}x,(JA_0')^{1/2}y\big) = (G'x,G'y)$$

and thus

$$\mathfrak{a}_1'[x,y] = (\mathfrak{a}_0'+\mathfrak{v})[x,y] = \big((1+C)G'x,G'y\big).$$

Consequently, noting that G' is self-adjoint in the Hilbert space $(\mathscr{K},(\cdot,\cdot))$,

$$JA_1 + \alpha\beta^{-1} + \delta = JA_1' = G'(1+C)G'.$$

Since, by assumption, $A_0 = A_0^{[*]}$ and also $A_0 = A_0^*$, we have $JA_0 = (JA_0)^* = A_0^*J = A_0J$, that is, J commutes with A_0. Hence J commutes with JA_0' and thus also with $(JA_0')^{1/2}$ and $(JA_0')^{-1/2}$, see [GGK90, Section V.4 ff.]. Let $z \in \rho(A_0)$. We obtain

$$A_1 - z = JG'\Big(1 - (\alpha\beta^{-1}+\delta)(JA_0')^{-1} - z(JA_0')^{-1/2}J(JA_0')^{-1/2} + C\Big)G'$$
$$= JG'\Big(1 - (\alpha\beta^{-1}+\delta+zJ)(JA_0')^{-1} + C\Big)G'.$$

Thus

$$(2.29) \qquad (A_1 - z)^{-1} = G'^{-1}\Big(1 - \big(\alpha\beta^{-1} + \delta + zJ\big)(JA_0')^{-1} + C\Big)^{-1} G'^{-1}J,$$

provided the second factor on the right exists and is bounded. This will be shown as follows. According to [Kat95, I-(4.24)] (the Neumann series), the factor

$$\big(1 - (\alpha\beta^{-1} + \delta + zJ)(JA_0')^{-1} + C\big)^{-1}$$

exists and is bounded if $\big(1 - (\alpha\beta^{-1} + \delta + zJ)(JA_0')^{-1}\big)^{-1}$ exists and has norm less than $\|C\|^{-1}$. Since $z \in \rho(A_0)$, the operator $A_0 - z$ is bijective, and hence also $JA_0 - zJ$. Consequently,

$$\begin{aligned}
1 - (\alpha\beta^{-1} + \delta + zJ)(JA_0')^{-1} &= \Big(JA_0' - (\alpha\beta^{-1} + \delta + zJ)\Big)(JA_0')^{-1} \\
&= \Big(J(A_0 + \alpha\beta^{-1}J + \delta J) - (\alpha\beta^{-1} + \delta + zJ)\Big)(JA_0')^{-1} \\
&= (JA_0 - zJ)(JA_0')^{-1}
\end{aligned}$$

is bijective. Since $\|C\| \le \beta$, it remains to show that $\beta M < 1$, where

$$M = \left\|\Big(1 - (\alpha\beta^{-1} + \delta + zJ)(JA_0')^{-1}\Big)^{-1}\right\|.$$

For M we have the estimate

$$\begin{aligned}
(2.30) \quad M &= \left\|\Big(1 - (\alpha\beta^{-1} + \delta + zJ)(JA_0')^{-1}\Big)^{-1}\right\| \\
&= \left\|JA_0'\Big(JA_0' - (\alpha\beta^{-1} + \delta + zJ)\Big)^{-1}\right\| \\
&= \left\|(JA_0 + \alpha\beta^{-1} + \delta)\Big((JA_0 + \alpha\beta^{-1} + \delta) - (\alpha\beta^{-1} + \delta + zJ)\Big)^{-1}\right\| \\
&= \left\|\big(\beta^{-1}(\alpha + \beta JA_0) + \delta\big)\big(J(A_0 - z)\big)^{-1}\right\| \\
&\le \beta^{-1}\big\|(\alpha + \beta JA_0)(A_0 - z)^{-1}\big\| + \delta\big\|(A_0 - z)^{-1}\big\| \\
&\le \beta^{-1}\Big(\alpha\big\|(A_0 - z)^{-1}\big\| + \beta\big\|A_0(A_0 - z)^{-1}\big\|\Big) + \delta\big\|(A_0 - z)^{-1}\big\|.
\end{aligned}$$

Now let $\delta = a(1 - a)(1 + a)^{-1}b^{-1}$ where

$$a = \alpha\big\|(A_0 - z)^{-1}\big\| + \beta\big\|A_0(A_0 - z)^{-1}\big\| \qquad \text{and} \qquad b = \beta\big\|(A_0 - z)^{-1}\big\|.$$

Since $a < 1$ by assumption (2.27), we have $\delta > 0$. Then, by (2.30), we have $M \leq \beta^{-1}a + \delta\beta^{-1}b$ and thus $\beta M < 1$ since

$$\beta M \leq a + \delta b = a + \frac{a(1-a)}{1+a} = \frac{2a}{1+a} < 1,$$

which is guaranteed by (2.27). Thus, according to [Kat95, I-(4.24)], the factor $\left(1 - \left(\alpha\beta^{-1} + \delta + zJ\right)(JA_0')^{-1} + C\right)^{-1}$ in (2.29) exists and is bounded.

A similar expression for $(A_0 - z)^{-1}$ is obtained by setting the linear operator $C = 0$ in (2.29):

(2.31) $\qquad (A_0 - z)^{-1} = G'^{-1}\left(1 - \left(\alpha\beta^{-1} + \delta + zJ\right)(JA_0')^{-1}\right)^{-1}G'^{-1}J.$

By [Kat95, I-(4.24)],

$$\left\|\left(1 - \left(\alpha\beta^{-1} + \delta + zJ + C\right)(JA_0')^{-1}\right)^{-1} - \left(1 - \left(\alpha\beta^{-1} + \delta + zJ\right)(JA_0')^{-1}\right)^{-1}\right\| \leq \frac{\beta M^2}{1 - \beta M},$$

for $\beta M < 1$, and hence inequalities (2.29) and (2.31) yield

(2.32) $\qquad \left\|(A_1 - z)^{-1} - (A_0 - z)^{-1}\right\| \leq \frac{\beta M^2}{(1 - \beta M)\delta}.$

Since

$$\beta M^2 = \frac{(\beta M)^2}{\beta} \leq \frac{4a^2}{\beta(1+a)^2}$$

and

$$(1 - \beta M)\delta \geq \left(1 - \frac{2a}{1+a}\right)\frac{a(1-a)}{b(1+a)} = \frac{a(1-a)^2}{b(1+a)^2},$$

inequality (2.28) follows from (2.32) and

$$\frac{\beta M^2}{(1 - \beta M)\delta} \leq \frac{4a^2}{\beta(1+a)^2} \cdot \frac{b(1+a)^2}{a(1-a)^2} = \frac{4ab}{\beta(1-a)^2} = \frac{4a\left\|(A_0 - z)^{-1}\right\|}{(1-a)^2}.$$

The case $\beta = 0$ can be dealt with by going to the limit $\beta \to 0$. ∎

Remark 2.40. In contrast to Theorem 2.32 it is not possible to choose $\alpha < 0$ in (2.27) of Theorem 2.39. This is due to inequality (2.30) which requires the stronger condition (2.27) compared to inequality (2.9) for the Hilbert space case.

Remark 2.41. Consider in Theorem 2.39 instead of A_1 the family of operators A_ε, $\varepsilon \in [0,1]$, which correspond to the sum of forms $\mathfrak{a}_\varepsilon := \mathfrak{a}_0 + \varepsilon \mathfrak{v}$, $\varepsilon \in [0,1]$, with $\mathfrak{a}_0$ and $\mathfrak{v}$ being the forms associated with A_0 and V_F, respectively. Then inequality (2.27) becomes (see proof of Theorem 2.39 above)

$$\left\| (A_\varepsilon - z)^{-1} - (A_0 - z)^{-1} \right\| < \varepsilon \frac{4 \Big(\alpha \left\| (A_0 - z)^{-1} \right\| + \beta \left\| A_0 (A_0 - z)^{-1} \right\| \Big) \left\| (A_0 - z)^{-1} \right\|}{\Big(1 - \big(\alpha \left\| (A_0 - z)^{-1} \right\| + \beta \left\| A_0 (A_0 - z)^{-1} \right\| \big) \Big)^2}.$$

Hence

$$\left\| (A_\varepsilon - z)^{-1} - (A_0 - z)^{-1} \right\| \longrightarrow 0, \quad \varepsilon \to 0,$$

and, by Theorem 1.27, A_ε converges to A_0 in the generalized sense.

Corollary 2.42. *Let A_0, V and A_1 be linear operators in a Krein space $(\mathcal{K}, [\cdot, \cdot])$ as in Theorem 2.39. Suppose that the spectrum of A_0 is separated into two parts $\sigma_1(A_0)$ and $\sigma_2(A_0)$ by a Cauchy contour Γ. If*

$$(2.33) \qquad \sup_{z \in \Gamma} \Big(\alpha \left\| (A_0 - z)^{-1} \right\| + \beta \left\| A_0 (A_0 - z)^{-1} \right\| \Big) < 1,$$

then the spectrum of A_1 is likewise separated into two parts $\sigma_1(A_1)$ and $\sigma_2(A_1)$, $\Gamma \subset \rho(A_1)$ and the results of Theorem 1.35 hold.

Proof. With Theorem 2.39 instead of Theorem 2.32 the proof is analogous to the proof of Corollary 2.34. ∎

The preceding results and Theorem 1.43 enable us to formulate Krein space versions of Theorems 2.35 and 2.36.

Theorem 2.43. *Let A_0 be a self-adjoint and V a symmetric operator in a Krein space $(\mathcal{K}, [\cdot, \cdot])$ such that A_0 and V are bounded from below in $(\mathcal{K}, [\cdot, \cdot])$ and A_0 is also self-adjoint in the Hilbert space $(\mathcal{K}, (\cdot, \cdot))$. Further, let V be relatively form-bounded with respect to A_0 with relative form-bounded less than $1/2$ and let A_ε, $\varepsilon \in [0,1]$, be the family of operators which correspond to the sum of forms $\mathfrak{a}_\varepsilon := \mathfrak{a}_0 + \varepsilon \mathfrak{v}$, $\varepsilon \in [0,1]$, with $\mathfrak{a}_0$ and $\mathfrak{v}$ being the forms associated with A_0 and V_F, respectively.*

(a) *Claims (i) and (ii) of Theorem 2.35 hold, respectively, if*

    (i) *the eigenvalue λ^0 is of definite type; in this case the eigenvalues contained in σ_1 are of the same type as λ^0.*

  (ii) *the eigenvalues λ_n^0, $n \in \mathbb{N}$, are of definite type; in this case the eigen-values contained in $\sigma_{n,\varepsilon}$ are of the same type as λ_n^0 for $n \in \mathbb{N}$ and $\varepsilon \in [0,\varepsilon_0]$.*

(b) *Claims (i) and (ii) of Theorem 2.36 hold, respectively, if*

  (i) *the spectral subspace $E(A_0,\sigma_0)\mathcal{K}$ is uniformly definite; in this case $E(A_\varepsilon,\sigma_\varepsilon)\mathcal{K}$ is of the same type as $E(A_0,\sigma_0)\mathcal{K}$ for $\varepsilon \in [0,1]$.*

  (ii) *the spectral subspace $E(A_0,\sigma_{n,0})\mathcal{K}$, $n \in \mathbb{N}$, is uniformly definite; in this case $E(A_\varepsilon,\sigma_{n,\varepsilon})\mathcal{K}$ is of the same type as $E(A_0,\sigma_{n,0})\mathcal{K}$ for $n \in \mathbb{N}$ and $\varepsilon \in [0,\varepsilon_0]$.*

Proof. We prove claim (i) of (a). The first part of the proof is analogous to the proof of Theorem 2.35. As in the proof of Theorem 2.35, let Γ be the positively oriented curve along the circle with center λ^0 and radius $\delta/2$. Then $\Gamma \subset \rho(A_0)$, $\{\lambda^0\} \subset \mathrm{int}(\Gamma)$ and $(\Gamma \cup \mathrm{int}(\Gamma)) \cap (\sigma(A_0)\backslash\{\lambda^0\}) = \emptyset$. By Proposition 1.38, since A_0 is also self-adjoint in the Hilbert space $(\mathcal{K},(\cdot,\cdot))$, inequality (2.33) is satisfied if (2.12) holds. By Corollary 2.42, the spectrum of A_ε is separated into the two parts $\sigma_\varepsilon := \mathrm{int}\,\Gamma \cap \sigma(A_\varepsilon)$ and $\sigma(A_\varepsilon)\backslash\sigma_\varepsilon$ such that σ_ε consists of isolated eigenvalues with total multiplicity m. By Theorem 2.39 and Remark 2.41, for one (and hence for all) $z \in \rho(A_\varepsilon)$ the resolvents of $A_\varepsilon = A_0 + \varepsilon V$, depend continuously on ε in the operator norm for $0 \le \varepsilon \le 1$. Let $n \in \mathbb{N}$. We apply Theorem 1.43. Since A_ε is self-adjoint in the Krein space $(\mathcal{K},[\cdot,\cdot])$, $\sigma(A_\varepsilon)$ is symmetric with respect to the real axis. Hence $\sigma_\varepsilon = \sigma_\varepsilon^*$ for all $0 \le \varepsilon \le 1$ since $\Gamma = \Gamma^*$. Since, by assumption, λ^0 is of definite type, $E(A_0,\sigma_0)\mathcal{K}$ is either uniformly positive or uniformly negative. According to Theorem 1.43 the subspace $E(A_\varepsilon,\sigma_\varepsilon)\mathcal{K}$ is of the same type as $E(A_0,\sigma_0)\mathcal{K}$ and the set σ_ε is real for all $0 \le \varepsilon \le 1$. Consequently, σ_ε consists of a finite system of real eigenvalues of A_ε with total multiplicity m which are of the same type as λ^0 for $0 \le \varepsilon \le 1$.

  Claim (ii) of (a) follows from (i) since the assumptions of (i) are satisfied for A_0 and εV with $\varepsilon \in [0,\varepsilon_0]$ for every eigenvalue λ_n^0, $n \in \mathbb{N}$, if we choose ε_0 such that

$$\varepsilon_0 \sup_{n \in \mathbb{N}} \left(\frac{1}{\delta_n}\left(\alpha_n + \beta_n\left(\delta_n + |\lambda_n^0|\right)\right)\right) < \frac{1}{2}.$$

  The proof of claims (i) and (ii) of (b) are analogous to those of (i) and (ii) of (a). ∎

2.3 Pseudo-Friedrichs Extensions

While the Friedrichs extension is defined for densely defined sectorial operators, this section introduces another kind of extension of the operator-sum $A_0 + V$, which can be applied to not necessarily sectorial operators. In contrast to the preceding paragraphs the main difference is the assumption $\mathscr{D}(V) \subset \mathscr{D}(A_0)$. The results of this section are not essentially related to sesquilinear forms, but the techniques used in the proofs are similar.

2.3.1 Perturbation of Spectra of Self-Adjoint Operators in Krein Spaces

The following theorem is a Krein space-generalisation of [Kat95, Theorem VI.3.11]. This result has been proved in a different way in [Ves72a], see Lemma 2.2 therein.

Theorem 2.44. *Let A_0 be a self-adjoint operator in a Krein space $\big(\mathscr{K}, [\cdot,\cdot]\big)$ and let V be an operator in $\big(\mathscr{K}, [\cdot,\cdot]\big)$ such that $\mathscr{D}(V) \subset \mathscr{D}(A_0)$ and*

$$(2.34) \qquad |[Vx,x]| \le \alpha\|x\|^2 + \beta\,[J|A_0|x,x], \quad x \in \mathscr{D}(V),$$

where $\alpha \ge 0$ and $0 \le \beta < 1$ or $0 \le \beta < 1/2$ according to whether V is symmetric in $\big(\mathscr{K}, [\cdot,\cdot]\big)$ or not. J denotes a fundamental symmetry on $\mathscr{K}$. If $\mathscr{D}(V)$ is a core of $|A_0|^{1/2}$, then there exists a unique closed extension A_1 of $A_0 + V$ such that $\mathscr{D}(A_1) \subset \mathscr{D}\big(|A_0|^{1/2}\big)$ and $\mathscr{D}(A_1^{[*]}) \subset \mathscr{D}\big(|A_0|^{1/2}\big)$. A_1 is self-adjoint in $\big(\mathscr{K}, [\cdot,\cdot]\big)$ if V is symmetric in $\big(\mathscr{K}, [\cdot,\cdot]\big)$.

Definition 2.45. The operator A_1 from Theorem 2.44 is called ***pseudo-Friedrichs extension of*** $A_0 + V$ in the Krein space $\big(\mathscr{K}, [\cdot,\cdot]\big)$.

Proof of Theorem 2.44. The proof follows from the Hilbert space version of the theorem, see [Kat95, Theorem VI.3.11]. Let $(\cdot,\cdot)$ denote a Hilbert space inner product on $\mathscr{K}$, i. e., $(\cdot,\cdot) = [J\cdot,\cdot]$. Then inequality (2.34) yields

$$|(JVx,x)| = |[Vx,x]| \le \alpha\|x\|^2 + \beta[J|A_0|x,x] = \alpha\|x\|^2 + \beta(|A_0|x,x), \quad x \in \mathscr{D}(V).$$

By assumption, $A_0 = A_0^{[*]}$ or, equivalently, $JA_0 = (JA_0)^*$. Thus

$$|A_0| = (A_0^*A_0)^{1/2} = (A_0^*J^2A_0)^{1/2} = \big((JA_0)^*(JA_0)\big)^{1/2} = |JA_0|.$$

Since $|A_0| = |JA_0|$ and the operators A_0 and JA_0 as well as V and JV have the same domains, respectively, [Kat95, Theorem VI.3.11] implies the existence of the unique pseudo-Friedrichs extension $\widetilde{A_1}$ of $JA_0 + JV$. That is,

$$(2.35) \qquad \mathcal{D}(\widetilde{A_1}) \supset \mathcal{D}(JA_0 + JV) \quad \text{and} \quad \widetilde{A_1}x = (JA_0 + JV)x,$$

for $x \in \mathcal{D}(JA_0 + JV) = \mathcal{D}(V)$. The operator $A_1 := J\widetilde{A_1}$ has domain $\mathcal{D}(A_1) = \mathcal{D}(J\widetilde{A_1}) = \mathcal{D}(\widetilde{A_1})$ and since $\mathcal{D}(JA_0 + JV) = \mathcal{D}(A_0 + V)$, (2.35) implies $\mathcal{D}(A_1) \supset \mathcal{D}(A_0 + V)$ and

$$A_1 = J\widetilde{A_1}x = J(JA_0 + JV)x = (A_0 + V)x, \quad x \in \mathcal{D}(A_0 + V).$$

That is, A_1 is the (unique) pseudo-Friedrichs extension of $A_0 + V$ in the Krein space $\mathcal{K}$. If V is symmetric with respect to $[\cdot, \cdot]$, then, by Lemma 1.8, JV is symmetric with respect to $(\cdot, \cdot)$. In this case, according to [Kat95, Theorem VI.3.11], $\widetilde{A_1}$ is self-adjoint with respect to $(\cdot, \cdot)$, and hence A_1 is self-adjoint with respect to the Krein space inner product $[\cdot, \cdot]$ by Lemma 1.8. ∎

Although the following result is rather convenient, it has has not explicitly been stated but proved in [Kat95]. It is a corollary to [Kat95, Theorem VI.3.11].

Lemma 2.46. *Let A_0 be a self-adjoint operator in a Hilbert space $(\mathcal{H}, (\cdot, \cdot))$ and let V be an operator such that $\mathcal{D}(V) \subset \mathcal{D}(A_0)$ and*

$$|(Vx, x)| \le \alpha \|x\|^2 + \beta (|A_0|x, x), \quad x \in \mathcal{D}(V),$$

where $\alpha \ge 0$ and $0 \le \beta < 1$ or $0 \le \beta < 1/2$ according to whether V is symmetric in $\mathcal{H}$ or not. Let $\mathcal{D}(V)$ be a core of $|A_0|^{1/2}$ and denote the pseudo-Friedrichs extension of $A_0 + V$ by A_1. Let $\kappa = 1$ if V is symmetric in $(\mathcal{H}, (\cdot, \cdot))$, otherwise $\kappa = 2$. If there is a point $z \in \rho(A_0)$ such that

$$(2.36) \qquad \kappa \left\| (\alpha + \beta |A_0|)(A_0 - z)^{-1} \right\| < 1,$$

then $z \in \rho(A_1)$ and

$$\left\| (A_1 - z)^{-1} - (A_0 - z)^{-1} \right\| < \frac{4\kappa \left\| (\alpha + \beta |A_0|)(A_0 - z)^{-1} \right\| \left\| (A_0 - z)^{-1} \right\|}{\left(1 - \kappa \left\| (\alpha + \beta |A_0|)(A_0 - z)^{-1} \right\| \right)^2}.$$

Proof. The claim follows from the proof of [Kat95, Theorem VI.3.11], compare inequality (3.24) therein. ∎

The following theorem gives a convenient criterion for spectral gaps between separated parts of the spectrum not to close; it has also been proved in [Ves08, Theorem 3.1].

Theorem 2.47. *Let A_0 be a self-adjoint and V a symmetric operator in a Hilbert space $(\mathcal{H}, (\cdot, \cdot))$ such that $\mathcal{D}(V) \subset \mathcal{D}(A_0)$ and*

$$|(Vx, x)| \le \alpha \|x\|^2 + \beta (|A_0|x, x), \quad x \in \mathcal{D}(V),$$

where $\alpha \ge 0$ and $0 \le \beta < 1$. Let $\mathcal{D}(V)$ be a core of $|A_0|^{1/2}$ and denote the pseudo-Friedrichs extension of $A_0 + V$ by A_1. Suppose the open interval (a, b) is a subset of $\rho(A_0)$. Define the interval I by

$$I := \left(a + (\alpha + \beta|a|), \, b - (\alpha + \beta|b|) \right).$$

If

$$\frac{1}{\delta} \left(\alpha + \beta \frac{|a| + |b|}{2} \right) < \frac{1}{2},$$

where $\delta = b - a$, then $I \ne \emptyset$ and I is a subset of $\rho(A_1)$.

Proof. The proof is analogous to that of Theorem 2.38. ∎

It is possible to formulate a Krein space version of Lemma 2.46:

Theorem 2.48. *Let A_0 and V be linear operators in a Krein space $(\mathcal{K}, [\cdot, \cdot])$ as in Theorem 2.44. Further, let A_0 be self-adjoint in the Hilbert space $(\mathcal{K}, (\cdot, \cdot))$. Denote the pseudo-Friedrichs extension of $A_0 + V$ by A_1. Let $\kappa = 1$ if V is symmetric in $(\mathcal{K}, [\cdot, \cdot])$, otherwise $\kappa = 2$. If there is a point $z \in \rho(A_0)$ such that*

$$(2.37) \qquad \kappa \left(\alpha \|(A_0 - z)^{-1}\| + \beta \||A_0|(A_0 - z)^{-1}\| \right) < 1,$$

then $z \in \rho(A_1)$ and

$$\|(A_1 - z)^{-1} - (A_0 - z)^{-1}\|$$

$$(2.38) \qquad < \frac{4\kappa \left(\alpha \|(A_0 - z)^{-1}\| + \beta \||A_0|(A_0 - z)^{-1}\| \right) \|(A_0 - z)^{-1}\|}{\left(1 - \kappa \left(\alpha \|(A_0 - z)^{-1}\| + \beta \||A_0|(A_0 - z)^{-1}\| \right) \right)^2}.$$

Proof. From [Kat95, VI-(3.21)] we obtain the identity

$$JA_1 = G'\bigl(JA_0(JA_0')^{-1} + C\bigr)G',$$

where $G' := (JA_0')^{1/2}$, $A_0' = |A_0| + \alpha\beta^{-1}J + \delta J$ with $\delta > 0$ to be determined later and C is a bounded linear operator in $\bigl(\mathcal{K}, [\cdot, \cdot]\bigr)$ with $\|C\| \le \kappa\beta$; note that since $A_0 = A_0^{[*]}$ or, equivalently, $JA_0 = (JA_0)^*$, we have $|JA_0| = |A_0|$. Since, further, $A_0 = A_0^*$, we have $JA_0 = (JA_0)^* = A_0^*J = A_0J$, that is, J commutes with A_0. Hence J commutes with JA_0' and thus also with $(JA_0')^{1/2}$ and $(JA_0')^{-1/2}$, see [GGK90, Section V.4 ff.]. Let $z \in \rho(A_0)$. Then

$$\begin{aligned}
A_1 - z &= JG'\bigl(JA_0(JA_0')^{-1} - z(JA_0')^{-1/2}J(JA_0')^{-1/2} + C\bigr)G' \\
&= JG'\bigl((JA_0 - zJ)(JA_0')^{-1} + C\bigr)G',
\end{aligned}$$

and thus

(2.39) $$(A_1 - z)^{-1} = G'^{-1}\Bigl(J(A_0 - z)(JA_0')^{-1} + C\Bigr)^{-1}G'^{-1}J,$$

provided the second factor on the right exists and is bounded. According to [Kat95, I-(4.24)] (the Neumann series), this is true if

(2.40) $$\left\|\Bigl(J(A_0 - z)(JA_0')^{-1}\Bigr)^{-1}\right\| < \frac{1}{\|C\|},$$

which is satisfied if

(2.41) $$\begin{aligned}
&\|C\|\left\|\Bigl(J(A_0 - z)(JA_0')^{-1}\Bigr)^{-1}\right\| \\
&\le \|C\|\left\|JA_0'(A_0 - z)^{-1}\right\| \\
&\le \kappa\beta\left\|(J|A_0| + \alpha\beta^{-1} + \delta)(A_0 - z)^{-1}\right\| \\
&\le \kappa\left\|(\alpha + \beta J|A_0|)(A_0 - z)^{-1}\right\| + \kappa\beta\delta\left\|(A_0 - z)^{-1}\right\| \\
&\le \kappa\Bigl(\alpha\left\|(A_0 - z)^{-1}\right\| + \beta\left\||A_0|(A_0 - z)^{-1}\right\|\Bigr) + \kappa\beta\delta\left\|(A_0 - z)^{-1}\right\| \\
&< 1.
\end{aligned}$$

Now let $\delta = a(1 - a)(1 + a)^{-1}b^{-1}$ where

$$a = \kappa\Bigl(\alpha\left\|(A_0 - z)^{-1}\right\| + \beta\left\||A_0|(A_0 - z)^{-1}\right\|\Bigr) \qquad \text{and} \qquad b = \kappa\beta\left\|(A_0 - z)^{-1}\right\|.$$

By assumption (2.37), $a < 1$ and thus $\delta > 0$. Then, by (2.41) and since

(2.42)

$$\kappa\Big(\alpha\big\|(A_0-z)^{-1}\big\| + \beta\big\||A_0|(A_0-z)^{-1}\big\|\Big) + \kappa\beta\delta\big\|(A_0-z)^{-1}\big\|$$

$$= a + \frac{a(1-a)}{1+a} = \frac{2a}{1+a},$$

inequality (2.40) is satisfied since $a < 1$ by (2.37). Equation (2.39) and a similar expression for $(A_0-z)^{-1}$ obtained by setting the linear operator $C = 0$ in (2.39) yield, by [Kat95, I-(4.24)],

(2.43) $$\big\|(A_1-z)^{-1} - (A_0-z)^{-1}\big\| \leq \frac{\kappa\beta\big\|JA_0'(A_0-z)^{-1}\big\|^2}{\Big(1 - \kappa\beta\big\|JA_0'(A_0-z)^{-1}\big\|\Big)\delta}.$$

Since, by (2.41) and (2.42),

$$\kappa\beta\big\|JA_0'(A_0-z)^{-1}\big\|^2 = \frac{\Big(\kappa\beta\big\|JA_0'(A_0-z)^{-1}\big\|\Big)^2}{\kappa\beta} \leq \frac{4a^2}{\kappa\beta(1+a)^2}$$

and

$$\Big(1 - \kappa\beta\big\|JA_0'(A_0-z)^{-1}\big\|\Big)\delta \geq \Big(1 - \frac{2a}{1+a}\Big)\frac{a(1-a)}{b(1+a)} = \frac{a(1-a)^2}{b(1+a)^2},$$

inequality (2.38) follows from (2.43) and

$$\frac{\kappa\beta\big\|JA_0'(A_0-z)^{-1}\big\|^2}{\Big(1 - \kappa\beta\big\|JA_0'(A_0-z)^{-1}\big\|\Big)\delta} \leq \frac{4a\big\|(A_0-z)^{-1}\big\|}{(1-a)^2}.$$

The case $\beta = 0$ can be dealt with by going to the limit $\beta \to 0$. ∎

Remark 2.49. Consider in Theorem 2.48 instead of A_1 the family of pseudo-Friedrichs extensions A_ε, $\varepsilon \in [0,1]$, of $A_0 + \varepsilon V$. Analogously to Remark 2.41, inequality (2.38) implies that A_ε converges to A_0 in the generalized sense.

Corollary 2.50. *Let A_0, V and A_1 be linear operators in a Krein space $(\mathcal{K}, [\cdot, \cdot])$ as in Theorem 2.48. Suppose that the spectrum of A_0 is separated into two parts $\sigma_1(A_0)$ and $\sigma_2(A_0)$ by a Cauchy contour Γ. If*

(2.44) $$\kappa\Big(\alpha\big\|(A_0-z)^{-1}\big\| + \beta\big\||A_0|(A_0-z)^{-1}\big\|\Big) < 1,$$

where $\kappa = 1$ or 2 according to whether V is symmetric in $\big(\mathscr{K},[\cdot,\cdot]\big)$ or not, then the spectrum of A_1 is likewise separated into two parts $\sigma_1(A_1)$ and $\sigma_2(A_1)$, $\Gamma \subset \rho(A_1)$ and the results of Theorem 1.35 hold.

Proof. With Theorem 2.48 the proof is analogous to the proof of Corollary 2.42. ∎

Now we are able to formulate the analogue of Theorem 2.43 for pseudo-Friedrichs extensions:

Theorem 2.51. *Let A_0 be a self-adjoint and V a symmetric operator in a Krein space $\big(\mathscr{K},[\cdot,\cdot]\big)$ such that $\mathscr{D}(V) \subset \mathscr{D}(A_0)$ and*

$$|[Vx,x]| \le \alpha \|x\|^2 + \beta\,[J|A_0|x,x], \quad x \in \mathscr{D}(V),$$

where $\alpha \ge 0$ and $0 \le \beta < 1/2$. Suppose that $\mathscr{D}(V)$ is a core of $|A_0|^{1/2}$ and A_0 is also self-adjoint in the Hilbert space $\big(\mathscr{K},(\cdot,\cdot)\big)$. Then the claims of Theorem 2.43 hold.

Proof. Using the preceding results, the proof is analogous to the proof of Theorem 2.43. ∎

Remark 2.52. A similar version of Theorem 2.51 has also been proved in [Ves72b], see Theorem 3 therein (compare also [Ves72a]). In comparison to Theorem 2.51, Theorem 3 in [Ves72b] requires further assumptions on the constants α, β and the length of the spectral gaps but shows that in addition to the reality of $\sigma(A_1)$ the pseudo-Friedrichs extension A_1 is similar to an operator which is self-adjoint in a Hilbert space, or, equivalently, A_1 is of scalar type, see [Ves72a], [Ves72b] and [DS88, Chapter XV].

Chapter 3

Examples

In this chapter we present examples and applications for the main results of Chapter 1 on relatively bounded perturbations of self-adjoint operators in Krein spaces. While the family of operators A_ε, $\varepsilon \in [0,1]$, of the first example (Section 3.1) is also self-adjoint in a Hilbert space, which includes the spectrum of A_ε being real, the operators A_ε of the remaining two examples (Section 3.2 and Section 3.3) are not. The family of operators considered in Section 3.3 was introduced by E. Caliceti and S. Graffi in [CG05].

3.1 Example 1

In the Hilbert space $L^2(-L,L)$, where $L \in \mathbb{R}_+$, besides the standard positive definite inner product

$$(f,g) = \int_{[-L,L]} f(x)\overline{g(x)}\,\mathrm{d}x, \quad f,g \in L^2(-L,L),$$

we consider the indefinite inner product defined by

$$(3.1) \qquad [f,g] := \int_{[-L,L]} f(x)\overline{g(-x)}\,\mathrm{d}x, \quad f,g \in L^2(-L,L).$$

Then $\left(L^2(-L,L),[\cdot,\cdot]\right)$ is a Krein space, a fundamental decomposition is given by

$$(3.2) \qquad L_e^2(-L,L)\,[\dotplus]\,L_o^2(-L,L),$$

where $L_e^2(-L,L)$ and $L_o^2(-L,L)$ are the sets of even and odd functions of $L^2(-L,L)$, respectively. We may give an explicit formula for the fundamental symmetry J in the Krein space $\left(L^2(-L,L),[\cdot,\cdot]\right)$ corresponding to decomposition (3.2). Each $f \in L^2(-L,L)$ can be decomposed as a sum $f = f_e + f_o$ with

$$f_e(x) := \frac{1}{2}\bigl(f(x)+f(-x)\bigr), \quad f_o(x) := \frac{1}{2}\bigl(f(x)-f(-x)\bigr),$$

where $f_e \in L^2_e(-L,L)$ and $f_o \in L^2_o(-L,L)$. For each such f we have

(3.3) $Jf(x) = f_e(x) - f_o(x) = f_e(-x) + f_o(-x) = f(-x)$.

Definition 3.1. In the Krein space $\left(L^2(-L,L),[\cdot,\cdot]\right)$, where $L \in \mathbb{R}_+$, we define the operators A_0 and V by

$$\mathscr{D}(A_0) := \{f \in L^2(-L,L) : f \in AC^2(-L,L), f'' \in L^2(-L,L), f(-L) = f(L) = 0\},$$

$$A_0 f := -\frac{d^2}{dx^2}f, \quad f \in \mathscr{D}(A_0),$$

and

$$\mathscr{D}(V) := \{f \in L^2(-L,L) : f \in AC(-L,L), f' \in L^2(-L,L), f(-L) = f(L) = 0\},$$

$$Vf := i\frac{d}{dx}f, \quad f \in \mathscr{D}(V).$$

We define the family of operators A_ε, $\varepsilon \in [0,1]$, by $A_\varepsilon := A_0 + \varepsilon V$, $\varepsilon \in [0,1]$.

Proposition 3.2. *The operator A_0 from Definition 3.1 is a self-adjoint operator in the Hilbert space $\left(L^2(-L,L),(\cdot,\cdot)\right)$ and in the Krein space $\left(L^2(-L,L),[\cdot,\cdot]\right)$.*

Proof. It is well-known that A_0 is self-adjoint in the Hilbert space $\left(L^2(-L,L), (\cdot,\cdot)\right)$, see [Kat95, Example V.3.25]. Hence, by equation (1.6) and by the definition of A_0, we have $A_0^{[*]}f(x) = JA_0Jf(x) = JA_0f(-x) = A_0f(x)$ for $f \in \mathscr{D}(A_0)$. By (3.3), $\mathscr{D}(A_0) = \mathscr{D}(JA_0J) = \mathscr{D}\left(A_0^{[*]}\right)$, that is, A_0 is a self-adjoint in the Krein space $\left(L^2(-L,L),[\cdot,\cdot]\right)$.   ∎

Proposition 3.3. *The operator V from Definition 3.1 is symmetric in the Krein space $\left(L^2(-L,L),[\cdot,\cdot]\right)$.*

Proof. It is well-known that V is symmetric in the Hilbert space $\left(L^2(-L,L), (\cdot,\cdot)\right)$, see [Kat95, Example V.3.14]. Thus, by equation (1.6),

$$V^{[*]}f(x) = JV^*Jf(x) = JVJf(x) = JVf(-x) = Vf(x), \quad f \in \mathscr{D}(V),$$

and hence V is symmetric in the Krein space $\left(L^2(-L,L),[\cdot,\cdot]\right)$.   ∎

Remark 3.4. It is a simple exercise to show that the spectrum of the unperturbed operator A_0 consists entirely of the eigenvalues:

$$\sigma(A_0) = \{\lambda_n^0 : n \in \mathbb{N}\}, \quad \lambda_n^0 := \left(\frac{n\pi}{2L}\right)^2, \quad n \in \mathbb{N}.$$

The distance of two consecutive eigenvalues is growing linearly with n:

$$\lambda_n^0 - \lambda_{n-1}^0 = \frac{n^2\pi^2}{4L^2} - \frac{(n-1)^2\pi^2}{4L^2} = \frac{(2n-1)\pi^2}{4L^2}, \quad n \in \mathbb{N}, n \geq 2.$$

Hence for $\delta_n = \mathrm{dist}(\lambda_n^0, \sigma(A_0)\setminus\{\lambda_n^0\})$, $n \in \mathbb{N}$, we establish $\delta_1 = \lambda_2^0 - \lambda_1^0$ and $\delta_n = \lambda_n^0 - \lambda_{n-1}^0$, $n \in \mathbb{N}$, $n \geq 2$.

Proposition 3.5. *For $[a,b] \subset \mathbb{R}$ and $n \in \mathbb{N}$ define*

$$\mathscr{D}(D) := \{f \in L^2(a,b) : f \in AC(a,b), f' \in L^2(a,b), f(a) = f(b)\},$$

$$Df := -\mathrm{i}\frac{\mathrm{d}}{\mathrm{d}x}f, \quad f \in \mathscr{D}(D).$$

Then, for every $n \in \mathbb{N}$, D^n is relatively bounded with respect to D^{2n} such that for any $\tau > 0$ we have

(3.4)
$$\|D^n f\|^2 \leq \frac{1}{4\tau}\|f\|^2 + \tau\|D^{2n}f\|^2, \quad f \in \mathscr{D}(D^{2n}).$$

Proof. Let $n \in \mathbb{N}$ and let $f \in \mathscr{D}(D^{2n})$. By the Cauchy-Schwarz inequality and since D^n is self-adjoint in the Hilbert space $(L^2(a,b), (\cdot,\cdot))$, we have

$$\|D^n f\|^2 = (D^n f, D^n f) = (D^{2n}f, f) \leq \|D^{2n}f\|\|f\| \leq \frac{1}{2\tilde{\tau}}\|f\|^2 + \frac{\tilde{\tau}}{2}\|D^{2n}f\|^2, \quad \tilde{\tau} > 0.$$

For $\tau := \tilde{\tau}/2$ we obtain the desired result. ∎

Remark 3.6. Proposition 3.5 is special case of a more general result; it is possible to show by induction that for every $m, n \in \mathbb{N}$, $m < n$, D^m is relatively compact with respect to D^n. Here we restrict us to the case $n = 2m$ because in the following we are particularly interested in the constants $1/(4\tau)$ and τ of (3.4).

Remark 3.7. Since A_0 and V are self-adjoint in the Hilbert space $(L^2(-L,L), (\cdot,\cdot))$, see Propositions 3.2 and 3.3, and since V is A_0-compact by Proposition 3.5, $A_1 = A_0 + V$ is self-adjoint in the Hilbert space $(L^2(-L,L), (\cdot,\cdot))$. Consequently, the spectrum of A_0 is real. Nevertheless, we show that also the assumptions of Theorem 1.44 (ii) are satisfied.

By Proposition 3.2, A_0 is self-adjoint in the Krein space $(L^2(-L,L), [\cdot,\cdot])$. According to Propositions 3.2, 3.3 and 3.5, A_ε is self-adjoint in $(L^2(-L,L), [\cdot,\cdot])$.

We have to check if condition (1.24) is satisfied. Set $\kappa := \pi/(2L)$. By Remark 3.4, we have $\delta_1 = 3\kappa^2$ and $\delta_n = (2n-1)\kappa^2$, $n \in \mathbb{N}$, $n \geq 2$. According to Proposition 3.5, we can choose $\alpha_1 = \kappa$ and $\beta_1 = 1/(4\kappa)$, and $\alpha_n = (\kappa\sqrt{n^2+2n-1})/2$ and $\beta_n = 1/(2\kappa\sqrt{n^2+2n-1})$, $n \in \mathbb{N}$, $n \geq 2$. Hence

$$
\gamma := \sup_{n\in\mathbb{N}}\left(\frac{1}{\delta_n}\left(\alpha_n + \beta_n\left(\delta_n + |\lambda_n^0|\right)\right)\right)
$$

$$
= \sup_{n\in\mathbb{N},\,n\geq2}\left(\frac{1}{(2n-1)\kappa^2}\left(\frac{\kappa\sqrt{n^2+2n-1}}{2} + \frac{1}{2\kappa\sqrt{n^2+2n-1}}\left((2n-1)\kappa^2+(n\kappa)^2\right)\right)\right)
$$

$$
= \sup_{n\in\mathbb{N},\,n\geq2}\left(\frac{\sqrt{n^2+2n-1}}{2\kappa(2n-1)} + \frac{1}{2\kappa\sqrt{n^2+2n-1}} + \frac{n^2}{2\kappa(2n-1)\sqrt{n^2+2n-1}}\right)
$$

$$
= \sup_{n\in\mathbb{N},\,n\geq2}\frac{\sqrt{n^2+2n-1}}{\kappa(2n-1)}
$$

$$
= \frac{\sqrt{7}}{3\kappa} < \infty,
$$

where we have used the fact that the function

$$
\frac{\sqrt{n^2+2n-1}}{\kappa(2n-1)}
$$

is monotonously decreasing for $n \geq 2$. If we choose $\varepsilon_0 = 1/(2\gamma)$, i.e.,

$$
\varepsilon_0 = \frac{3\kappa}{2\sqrt{7}} = \frac{3\pi}{4\sqrt{7}L},
$$

then (1.24) holds for all $\varepsilon \in [0, \varepsilon_0) \cap [0,1]$, where $\varepsilon_0 > 1$ is possible.

In fact, the eigenvalues of A_1 are obtained by solving the differential equation

$$
(3.5) \qquad \frac{d^2}{dx^2}f - i\frac{d}{dx}f + \lambda f = 0
$$

on the interval $[-L, L]$ with the boundary conditions given by Definition 3.1. The general solution of (3.5) is

$$
(3.6) \qquad f(x) := \gamma_1 e^{\frac{i}{2}\left(1-\sqrt{4\lambda+1}\right)x} + \gamma_2 e^{\frac{i}{2}\left(1+\sqrt{4\lambda+1}\right)x},
$$

for $x \in [-L,L]$ and $\lambda \in \mathbb{C}$, with arbitrary constants γ_i, $i = 1,2$. Substituting the boundary conditions $f(-L) = f(L) = 0$ into (3.6), we obtain the system of equations

$$\begin{pmatrix} e^{-\frac{i}{2}L\left(1-\sqrt{4\lambda+1}\right)} & e^{-\frac{i}{2}L\left(1+\sqrt{4\lambda+1}\right)} \\ e^{\frac{i}{2}L\left(1-\sqrt{4\lambda+1}\right)} & e^{\frac{i}{2}L\left(1+\sqrt{4\lambda+1}\right)} \end{pmatrix} \begin{pmatrix} \gamma_1 \\ \gamma_2 \end{pmatrix} = \begin{pmatrix} 0 \\ 0 \end{pmatrix}$$

which leads to non-trivial solutions if the determinant equals 0:

$$2i\sin\left(L\sqrt{4\lambda+1}\right) = 0 \iff \lambda = \left(\frac{n\pi}{2L}\right)^2 - \frac{1}{4}, \quad n \in \mathbb{N}_0.$$

If $\lambda = -1/4$, then f is the trivial solution, and thus the eigenvalues of A_1 are

$$\lambda_n^0 = \left(\frac{n\pi}{2L}\right)^2 - \frac{1}{4}, \quad n \in \mathbb{N}.$$

3.2 Example 2

Definition 3.8. In the Krein space $\left(L^2(-L,L),[\cdot,\cdot]\right)$, where $L \in \mathbb{R}_+$ and $[\cdot,\cdot]$ as in (3.1), we define the operators A_0 and V by

$$\mathscr{D}(A_0) := \left\{f \in L^2(-L,L): f \in AC^4(-L,L), f^{(4)} \in L^2(-L,L),\right.$$

$$f(-L) = f(L) = f''(-L) = f''(L) = 0\Big\},$$

$$A_0 f := \frac{d^4}{dx^4}f, \quad f \in \mathscr{D}(A_0),$$

and

$$\mathscr{D}(V) := \left\{f \in L^2(-L,L): f \in AC^2(-L,L), f'' \in L^2(-L,L), f(-L) = f(L) = 0\right\},$$

$$Vf := i\frac{d}{dx}\left(h(x)\frac{d}{dx}f\right), \quad f \in \mathscr{D}(V),$$

where $h \in C^2(-L,L)$ is a real-valued function having the following properties:

$$h(-x) = -h(x) \quad \text{for } x \in [-L,L], \qquad h(x) > 0 \quad \text{for } x \in (0,L).$$

We define the family of operators $A_\varepsilon := A_0 + \varepsilon V$, $\varepsilon \in [0,1]$.

Proposition 3.9. *The operator A_0 from Definition 3.8 is a self-adjoint operator in the Hilbert space $\left(L^2(-L,L),(\cdot,\cdot)\right)$ and in the Krein space $\left(L^2(-L,L),[\cdot,\cdot]\right)$.*

Proof. The calculations are similar to those of the preceding example, compare Proposition 3.2. ∎

Proposition 3.10. *The operator V from Definition 3.8 is symmetric in the Krein space $(L^2(-L,L),[\cdot,\cdot])$.*

Proof. Integration by parts yields for $f,g \in \mathscr{D}(V)$:

$$
\begin{aligned}
[Vf,g] &= \int_{[-L,L]} i\frac{d}{dx}\left(h(x)\frac{d}{dx}f(x)\right)\overline{g(-x)}\,dx \\
&= -i\int_{[-L,L]} h(x)\frac{d}{dx}f(x)\overline{\frac{d}{dx}g(-x)}\,dx \\
&= i\int_{[-L,L]} \frac{d}{dx}f(x)\overline{h(-x)\frac{d}{dx}g(-x)}\,dx \\
&= \int_{[-L,L]} f(x)\overline{i\frac{d}{dx}\left(h(-x)\frac{d}{dx}g(-x)\right)}\,dx \\
&= [f,Vg].
\end{aligned}
$$
∎

Proposition 3.11. *Let A_0 be the operator defined in Definition 3.8. Then the spectrum of A_0 consists of the eigenvalues*

$$
\lambda_n^0 := \left(\frac{n\pi}{2L}\right)^4, \quad n \in \mathbb{N}.
$$

Proof. Since A_0 is the square of A_0 from Definition 3.1 of Example 1, the statement follows from the spectral mapping theorem, see, e.g., [GGK90, Theorem 3.3]. ∎

Remark 3.12. For the distance of two consecutive eigenvalues of A_0, we have:

$$
(3.7)\quad \lambda_n^0 - \lambda_{n-1}^0 = \frac{n^4\pi^4}{(2L)^4} - \frac{(n-1)^4\pi^4}{(2L)^4} = \frac{(4n^3 - 6n^2 + 4n - 1)\pi^4}{(2L)^4}, \quad n \in \mathbb{N}, n \geq 2.
$$

Remark 3.13. Let A_0 and V be linear operators as in Definition 3.8. Then V is relatively bounded with respect to A_0 with relative bound 0 such that for every $\tau > 0$ inequality (1.7) holds with $\alpha_\tau = c/\tau$ and $\beta_\tau = \tau$, where $c > 0$ is constant.

Proof. Let $f \in \mathscr{D}(A_0) \subset \mathscr{D}(V)$. Since

$$Vf = ih'(x)\frac{\mathrm{d}}{\mathrm{d}x}f + ih(x)\frac{\mathrm{d}^2}{\mathrm{d}x^2}f,$$

we have

$$\|Vf\| \le \sup|h'|\left\|\frac{\mathrm{d}}{\mathrm{d}x}f\right\| + \sup|h|\left\|\frac{\mathrm{d}^2}{\mathrm{d}x^2}f\right\|$$

and hence, by Proposition 3.5, we obtain the desired result. ■

Theorem 3.14. *Consider the family of operators A_ε, $\varepsilon \in [0,1]$, from Definition 3.1 in the Krein space $\left(L^2(-L,L),[\cdot,\cdot]\right)$. Then there exists an $\varepsilon_0 \in (0,1]$ such that for all $\varepsilon \in [0,\varepsilon_0)$ the spectrum of $A_\varepsilon = A_0 + \varepsilon V$ consists of simple and real eigenvalues and which are of definite type.*

Proof. By Proposition 3.9, A_0 is self-adjoint in the Hilbert space $\left(L^2(-L,L), (\cdot,\cdot)\right)$ and in Krein space $\left(L^2(-L,L),[\cdot,\cdot]\right)$. According to Proposition 3.10 and Remark 3.13, V is symmetric in $\left(L^2(-L,L),[\cdot,\cdot]\right)$ and A_0-bounded with A_0-bound 0. To complete the proof we have to check if condition (1.24) is satisfied.

The distance $\delta_n := \mathrm{dist}\left(\lambda_n^0, \sigma(A_0)\backslash\{\lambda_n^0\}\right)$ equals the distance of two consecutive eigenvalues. Set $\kappa := \pi/(2L)$. By equation (3.7), we have $\delta_1 = 15\kappa^4$ and $\delta_n = \left(4n^3 - 6n^2 + 4n - 1\right)\kappa^4$, $n \in \mathbb{N}$, $n \ge 2$. By Remark 3.13, we can choose $\alpha_1 = 4\kappa^2$ and $\beta_1 = c/(4\kappa^2)$, and

$$\alpha_n = \kappa^2\sqrt{c\left(n^4 + 4n^3 - 6n^2 + 4n - 1\right)} \quad \text{and} \quad \beta_n = \frac{c}{\kappa^2\sqrt{c\left(n^4 + 4n^3 - 6n^2 + 4n - 1\right)}}$$

for $n \in \mathbb{N}$, $n \ge 2$. Hence

$$\gamma := \sup_{n \in \mathbb{N}}\left(\frac{1}{\delta_n}\left(\alpha_n + \beta_n\left(\delta_n + |\lambda_n^0|\right)\right)\right)$$

$$= \max\left\{\frac{4(c+1)}{15\kappa^2}, \sup_{n \in \mathbb{N}, n\ge 2}\frac{2\sqrt{c\left(n^4 + 4n^3 - 6n^2 + 4n - 1\right)}}{\kappa^2(4n^3 - 6n^2 + 4n - 1)}\right\}$$

$$= \max\left\{\frac{4(c+1)}{15\kappa^2}, \frac{2\sqrt{31c}}{15\kappa^2}\right\} < \infty,$$

where we have used the fact that the function

$$\frac{2\sqrt{c\left(n^4 + 4n^3 - 6n^2 + 4n - 1\right)}}{\kappa^2\left(4n^3 - 6n^2 + 4n - 1\right)}$$

is monotonously decreasing for $n \geq 2$. If we choose $\varepsilon_0 = 1/(2\gamma)$, i.e.,

$$\varepsilon_0 = \min\left\{\frac{15\kappa^2}{8(c+1)}, \frac{15\kappa^2}{4\sqrt{31c}}\right\} = \min\left\{\frac{15\pi^2}{32L^2(c+1)}, \frac{15\pi^2}{16L^2\sqrt{31c}}\right\},$$

then (1.24) holds for all $\varepsilon \in [0, \varepsilon_0) \cap [0, 1]$, where $\varepsilon_0 > 1$ is possible. Now the claim follows from Theorem 1.44 (ii). ∎

3.3 A Class of Schrödinger Operators with Relatively Bounded Complex Potentials and Real Spectrum

In this paragraph we consider operators induced by the differential expression

$$A_\varepsilon = -\frac{\mathrm{d}^2}{\mathrm{d}x^2} + V_P + \varepsilon V_Q, \quad \varepsilon \in [0, 1],$$

in the Krein space $\left(L^2(\mathbb{R}), [\cdot, \cdot]\right)$. Here V_P and V_Q are multiplication operators with $P(x)$ and $\mathrm{i}Q(x)$, respectively, where $P(x)$ is a real, even polynomial of degree $2p$, $p \geq 1$, with $\lim_{|x|\to\infty} P(x) = \infty$, and $Q(x)$ is a real, odd polynomial of degree $2q - 1$, $q \geq 1$, such that $p > 2q$.

This class of operators was considered in [CG05], where the results were proved by means of perturbation theory for linear operators. Here we are able to show, it is also an example for the results of Chapter 1.

Definition 3.15. Consider the Krein space $\left(L^2(\mathbb{R}), [\cdot, \cdot]\right)$, with $[\cdot, \cdot]$ given by

$$(3.8) \qquad [f, g] := \int_{\mathbb{R}} f(x)\overline{g(-x)}\,\mathrm{d}x, \quad f, g \in L^2(\mathbb{R}).$$

Let $P : \mathbb{R} \to \mathbb{R}$ be an even polynomial of degree $2p$, $p \geq 1$, such that

$$\lim_{|x|\to\infty} P(x) = \infty$$

and let $Q : \mathbb{R} \to \mathbb{R}$ be an odd polynomial of degree $2q - 1$, $q \geq 1$, such that $p > 2q$. Then we define the multiplication operators V_P and V_Q by

$$\mathscr{D}(V_P) := \{f \in L^2(\mathbb{R}) : Pf \in L^2(\mathbb{R})\},$$
$$V_P f(x) := P(x)f(x), \quad f \in \mathscr{D}(V_P),$$

and

$$\mathscr{D}(V_Q) := \{f \in L^2(\mathbb{R}) : iQf \in L^2(\mathbb{R})\},$$
$$V_Q f(x) := iQ(x)f(x), \quad f \in \mathscr{D}(V_Q).$$

Furthermore, we define the family of operators A_ε, $\varepsilon \in [0,1]$,

$$\mathscr{D}(A_\varepsilon) := \{f \in L^2(\mathbb{R}) : f \in AC^2(\mathbb{R}), -f'' + V_P f + \varepsilon V_Q f \in L^2(\mathbb{R})\},$$
$$A_\varepsilon f := \left(-\frac{d^2}{dx^2} + V_P + \varepsilon V_Q\right)f, \quad f \in \mathscr{D}(A_\varepsilon), \quad \varepsilon \in [0,1].$$

Proposition 3.16. *The operator A_0 defined in Definition 3.15 is self-adjoint in the Hilbert space $(L^2(\mathbb{R}),(\cdot,\cdot))$ and self-adjoint in the Krein space $(L^2(\mathbb{R}),[\cdot,\cdot])$. (Here $(\cdot,\cdot)$ denotes the standard positive definite Hilbert space inner product on $L^2(\mathbb{R})$).*

Proof. By [BS91, Theorem 2.1.1], the minimal operator $_0A_0 := A_0|_{C_0^\infty(\mathbb{R})}$ is essentially self-adjoint in $(L^2(\mathbb{R}),(\cdot,\cdot))$. Since $_0A_0 \subset \overline{_0A_0} \subset A_0$, we have $A_0^* \subset \overline{_0A_0}^* = _0A_0 \subset A_0$ and thus A_0 is self-adjoint in $(L^2(\mathbb{R}),(\cdot,\cdot))$. Using the fact that $P(x)$ is an even polynomial, integration by parts yields for $f,g \in \mathscr{D}(A_0)$:

$$[A_0 f, g] = \int_{\mathbb{R}} \left(-\frac{d^2}{dx^2} + V_P\right) f(x)\overline{g(-x)}\,dx$$
$$= \int_{\mathbb{R}} -\frac{d^2}{dx^2} f(x)\overline{g(-x)}\,dx + \int_{\mathbb{R}} P(x)f(x)\overline{g(-x)}\,dx$$
$$= \int_{\mathbb{R}} -f(x)\overline{\frac{d^2}{dx^2}g(-x)}\,dx + \int_{\mathbb{R}} f(x)\overline{P(-x)g(-x)}\,dx$$
$$= \int_{\mathbb{R}} f(x)\overline{\left(-\frac{d^2}{dx^2} + V_P\right)g(-x)}\,dx$$
$$= [f, A_0 g].$$

Hence A_0 is also symmetric in the Krein space $(L^2(\mathbb{R}),[\cdot,\cdot])$ and, since $\mathscr{D}(A_0) = \mathscr{D}(JA_0 J) = \mathscr{D}(A_0^{[*]})$, A_0 is even self-adjoint in $(L^2(\mathbb{R}),[\cdot,\cdot])$. ∎

The following remark and its proof can be found in [BS91, Theorem 2.3.1] and [CG05, (1.3)].

Remark 3.17. The family of operators A_ε, $\varepsilon \in [0,1]$, defined in Definition 3.15 has discrete spectrum. Moreover, the eigenvalues λ_n^0, $n \in \mathbb{N}$, of the unperturbed operator A_0 are simple, form an increasing sequence and satisfy the estimate

$$(3.9) \qquad \lambda_n^0 = c_1 n^{\frac{2p}{p+1}} + \mathcal{O}\left(n^{\frac{p-1}{p+1}}\right), \quad n \to \infty,$$

with some constant $c_1 > 0$; here $\mathcal{O}$ denotes the Landau symbol.

The following proposition can be found in [CG05, Lemma 2.1].

Proposition 3.18. *The multiplication operator V_Q given by Definition 3.15 is relatively bounded with respect to A_0 with A_0-bound 0. Moreover we have the following. For all $n \in \mathbb{N}$ there exist constants $c_2, c_3 > 0$ such that*

$$(3.10) \qquad \|V_Q f\| \le c_2 n^{\frac{p-1}{p+1}} \|f\| + \frac{c_3}{n} \|A_0 f\|, \quad f \in \mathscr{D}(A_0).$$

Proof. According to [Sim70], there exist $\gamma_1, \gamma_2 > 0$ such that

$$(3.11) \qquad \left\| \frac{d^2}{dx^2} f \right\| + \|V_P f\| \le \gamma_1 \|f\| + \gamma_2 \|A_0 f\|, \quad f \in \mathscr{D}(A_0).$$

Therefore it is sufficient to show that

$$(3.12) \qquad \|V_Q f\| \le \widetilde{\alpha}_n \|f\| + \widetilde{\beta}_n \|V_P f\|, \quad f \in \mathscr{D}(V_P),$$

where $\widetilde{\alpha}_n = \gamma_3 n^{\frac{p-1}{p+1}}$ and $\widetilde{\beta}_n = \gamma_4/n$ for $n \in \mathbb{N}$, $\gamma_3, \gamma_4 > 0$, since then, by (3.11),

$$\|V_Q f\| \le \left(\widetilde{\alpha}_n + \widetilde{\beta}_n \gamma_1 \right) \|f\| + \widetilde{\beta}_n \gamma_2 \|A_0 f\|, \quad f \in \mathscr{D}(V_P),$$

which implies (3.10). By Lemma 1.13, inequality (3.12) is equivalent to

$$(3.13) \qquad \|V_Q f\|^2 \le \widetilde{\alpha}_n^2 \|f\|^2 + \widetilde{\beta}_n^2 \|V_P f\|^2, \quad f \in \mathscr{D}(V_P).$$

Inequality (3.13) is implied by the pointwise inequality

$$(3.14) \qquad 0 \le \widetilde{\alpha}_n^2 + \widetilde{\beta}_n^2 P(x)^2 - Q(x)^2, \quad x \in \mathbb{R},$$

since (3.14) leads to

$$0 \le \int_{\mathbb{R}} \left(\widetilde{\alpha}_n^2 + \widetilde{\beta}_n^2 P(x)^2 - Q(x)^2 \right) |f(x)|^2 dx = \widetilde{\alpha}_n^2 \|f\|^2 - \|V_Q f\|^2 + \widetilde{\beta}_n^2 \|V_P f\|^2,$$

for $f \in L^2(\mathbb{R})$. $P(x)$ and $Q(x)$ can be minorized and majorized, respectively, by homogeneous polynomials of degree $2p$ and $2q - 1$, respectively. Thus, up to an additive constant, which can be absorbed in the constant γ_1 in (3.11), we can restrict ourselves to verify (3.14) for homogeneous polynomials $P(x)$ and $Q(x)$ of degree $2p$ and $2q-1$, respectively. Let $\widetilde{\beta}_n = \gamma_4/n$, $n \in \mathbb{N}$, for some $\gamma_4 > 0$. Then (3.14) is satisfied if

$$(3.15) \qquad 0 \le \widetilde{\alpha}_n^2 + \frac{\gamma_4^2}{n^2} x^{4p} - x^{4q-2}, \quad x \in \mathbb{R}.$$

Since there are only even powers, we can restrict ourselves to $x \ge 0$. Set

$$s := \frac{1}{2(p-q)+1}.$$

We consider two cases. First, let $x \ge n^s$. Hence

$$x \ge n^s \iff x^{4q-2} \le n^{-2} x^{4p} \iff 0 \le n^{-2} x^{4p} - x^{4q-2}, \quad x \in \mathbb{R}_0^+,$$

and thus (3.15) is satisfied with $\widetilde{\alpha}_n = 0$, $n \in \mathbb{N}$, and $\gamma_4 = 1$.

Now let $x < n^s$. Then we have $x^{4q-2} < n^{(4q-2)s}$ and thus (3.15) is satisfied with

$$\widetilde{\alpha}_n = n^{\frac{2q-1}{2(p-q)+1}} \quad \text{and} \quad \gamma_4 = 0.$$

By Definition 3.15, $p > 2q$ which implies

$$\frac{2q-1}{2(p-q)+1} < \frac{p-1}{p+1}. \qquad \blacksquare$$

Proposition 3.19. *The multiplication operator V_Q defined in Definition 3.15 is symmetric in the Krein space $\left(L^2(\mathbb{R}), [\cdot, \cdot]\right)$.*

Proof. Since $Q(x)$ is a real, odd polynomial, we have

$$\overline{iQ(-x)} = -iQ(-x) = iQ(x), \quad x \in \mathbb{R}.$$

Hence V is symmetric in $\left(L^2(\mathbb{R}), [\cdot, \cdot]\right)$:

$$[V_Q f, g] = \int_{\mathbb{R}} V_Q f(x) \overline{g(-x)} \mathrm{d}x = \int_{\mathbb{R}} iQ(x) f(x) \overline{g(-x)} \mathrm{d}x$$
$$= \int_{\mathbb{R}} f(x) \overline{iQ(-x)} g(-x) \mathrm{d}x = \int_{\mathbb{R}} f(x) \overline{V_Q g(-x)} \mathrm{d}x = [f, V_Q g],$$

for all $f, g \in \mathscr{D}(V_Q)$. $\qquad \blacksquare$

Theorem 3.20. *The family of operators A_ε, $\varepsilon \in [0,1]$, defined in Definition 3.15 is self-adjoint in the Krein space $(L^2(\mathbb{R}), [\cdot,\cdot])$ for all $\varepsilon \in [0,1]$, and there exists an $\varepsilon_0 \in [0,1]$ such that for all $\varepsilon \in [0,\varepsilon_0)$ the spectrum of $A_\varepsilon = A_0 + \varepsilon V$ consists of simple real eigenvalues of definite type.*

Proof. By Propositions 3.19 and 3.18, V is symmetric and A_0-bounded with A_0-bound 0. Hence, since A_0 is self-adjoint in $(L^2(\mathbb{R}), [\cdot,\cdot])$ by Proposition 3.16, $A_\varepsilon = A_0 + \varepsilon V$ is self-adjoint for all $\varepsilon \in [0,1]$ by Theorem 1.24. To complete the proof we have to check if condition (1.24) is satisfied.

The distance $\delta_n := \mathrm{dist}(\lambda_n^0, \sigma(A_0) \setminus \{\lambda_n^0\})$ is the minimum of the distances between the eigenvalues λ_n^0 and λ_{n-1}^0, or λ_n^0 and λ_{n+1}^0, i.e.,

$$\delta_n = \min\{\lambda_n^0 - \lambda_{n-1}^0, \lambda_{n+1}^0 - \lambda_n^0\}, \quad n \in \mathbb{N}, n \geq 2.$$

Since

$$n^{\frac{2p}{p+1}} - (n-1)^{\frac{2p}{p+1}} = n \cdot n^{\frac{p-1}{p+1}} - (n-1)(n-1)^{\frac{p-1}{p+1}}$$

$$= n^{\frac{p-1}{p+1}}\left(n - (n-1)\left(\frac{n-1}{n}\right)^{\frac{p-1}{p+1}}\right)$$

$$= n^{\frac{p-1}{p+1}}\left(n - n\left(\frac{n-1}{n}\right)^{\frac{p-1}{p+1}} + \left(\frac{n-1}{n}\right)^{\frac{p-1}{p+1}}\right)$$

$$= n^{\frac{p-1}{p+1}}\left(\frac{1-\left(\frac{n-1}{n}\right)^{\frac{p-1}{p+1}}}{n^{-1}} + \left(\frac{n-1}{n}\right)^{\frac{p-1}{p+1}}\right),$$

we have

$$\lambda_n^0 - \lambda_{n-1}^0 = c_1\left(n^{\frac{2p}{p+1}} - (n-1)^{\frac{2p}{p+1}}\right) + \mathcal{O}\left(n^{\frac{p-1}{p+1}}\right) - \mathcal{O}\left((n-1)^{\frac{p-1}{p+1}}\right)$$

$$= c_1 n^{\frac{p-1}{p+1}}\left(\frac{1-\left(\frac{n-1}{n}\right)^{\frac{p-1}{p+1}}}{n^{-1}} + \left(\frac{n-1}{n}\right)^{\frac{p-1}{p+1}}\right) + \mathcal{O}\left(n^{\frac{p-1}{p+1}}\right), \quad n \to \infty,$$

and thus

$$\frac{\lambda_n^0 - \lambda_{n-1}^0}{n^{\frac{p-1}{p+1}}} = c_1\left(\frac{1-\left(\frac{n-1}{n}\right)^{\frac{p-1}{p+1}}}{n^{-1}} + \left(\frac{n-1}{n}\right)^{\frac{p-1}{p+1}}\right) + \mathcal{O}(1), \quad n \to \infty.$$

By l'Hospital's rule,

$$\lim_{n \to \infty} \frac{1 - \left(\frac{n-1}{n}\right)^{\frac{p-1}{p+1}}}{n^{-1}} = \lim_{n \to \infty} \frac{-\frac{p-1}{p+1}\left(\frac{n-1}{n}\right)^{\frac{p-1}{p+1}-1}\frac{1}{n^2}}{-n^{-2}}$$

$$= \lim_{n \to \infty} \frac{p-1}{p+1}\left(\frac{n-1}{n}\right)^{\frac{-2}{p+1}}$$

$$= \frac{p-1}{p+1},$$

and we obtain

$$\limsup_{n \to \infty} \frac{\lambda_n^0 - \lambda_{n-1}^0}{n^{\frac{p-1}{p+1}}} < \infty, \quad \limsup_{n \to \infty} \frac{n^{\frac{p-1}{p+1}}}{\lambda_n^0 - \lambda_{n-1}^0} < \infty.$$

Consequently,

$$\delta_n \sim c_4 n^{\frac{p-1}{p+1}}, \quad \frac{\lambda_n^0}{\delta_n} \sim c_5 n, \quad n \to \infty,$$

with constants $c_4, c_5 > 0$, where $\sim$ stands for asymptotic equivalence (two sequences $(s_n)_{n=1}^\infty$ and $(t_n)_{n=1}^\infty$ are called asymptotically equivalent, in symbols, $s_n \sim t_n$, if and only if $s_n \in \mathcal{O}(t_n)$ and $t_n \in \mathcal{O}(s_n)$). If we set $\alpha_n := c_2 n^{\frac{p-1}{p+1}}$ and $\beta_n := c_3/n$, $n \in \mathbb{N}$, with $c_2, c_3 > 0$ according to Proposition 3.18, then, by Proposition 3.18,

$$\sup_{n \in \mathbb{N}}\left(\frac{1}{\delta_n}\left(\alpha_n + \beta_n\left(\delta_n + |\lambda_n^0|\right)\right)\right) < \infty.$$

Thus there exists an $\varepsilon_0 \in [0, 1]$ such that (1.24) holds for all $\varepsilon \in [0, \varepsilon_0)$, and the claim follows from Theorem 1.44 (ii). ∎

Remark 3.21. The last theorem has also been proved in works by E. Caliceti and S. Graffi (see [CG05, Theorem 1.1], compare also [Cal04], [CGS05], [Cal05] and [CCG06]).

Remark 3.22. Similar operators with different assumptions on the polynomials P and Q are considered in works by Dorey et al. (see [DDT01b]) and Shin (see [Shi02]).

Bibliography

[AI89]   T. Ya. Azizov and I. S. Iokhvidov, *Linear operators in spaces with an indefinite metric*, Pure and Applied Mathematics (New York), John Wiley & Sons Ltd., Chichester, 1989, Translated from the Russian by E. R. Dawson, A Wiley-Interscience Publication.

[AK04]   S. Albeverio and S. Kuzhel, *Pseudo-Hermiticity and theory of singular perturbations*, Lett. Math. Phys. **67** (2004), no. 3, 223–238.

[And79]   T. Ando, *Linear operators on Kreĭn spaces*, Hokkaido University Research Institute of Applied Electricity Division of Applied Mathematics, Sapporo, 1979.

[AT10]   T. Y. Azizov and C. Trunk, *On domains of $\mathscr{PT}$ symmetric operators related to $-y''(x)+(-1)^n x^{2n} y(x)$*, J. Phys. A **43** (2010), no. 17, 175303.

[BB98]   C. M. Bender and S. Boettcher, *Real spectra in non-Hermitian Hamiltonians having $\mathscr{PT}$ symmetry*, Phys. Rev. Lett. **80** (1998), no. 24, 5243–5246.

[BBJ03]   C. M. Bender, D. C. Brody, and H. F. Jones, *Must a Hamiltonian be Hermitian?*, Amer. J. Phys. **71** (2003), no. 11, 1095–1102.

[BBM99]   C. M. Bender, S. Boettcher, and P. N. Meisinger, *$\mathscr{PT}$-symmetric quantum mechanics*, J. Math. Phys. **40** (1999), no. 5, 2201–2229.

[Ben04a]   C. M. Bender, *Complex extension of quantum mechanics*, Symmetry in nonlinear mathematical physics. Part 1, 2, 3, Pr. Inst. Mat. Nats. Akad. Nauk Ukr. Mat. Zastos., 50, Part 1, vol. 2, Natsīonal. Akad. Nauk Ukraïni Īnst. Mat., Kiev, 2004, pp. 617–628.

[Ben04b]   _____ , *$\mathscr{PT}$ symmetry in quantum field theory*, Czechoslovak J. Phys. **54** (2004), no. 1, 13–28, Pseudo-Hermitian Hamiltonians in quantum physics.

[Ben07]   _____ , *Making sense of non-Hermitian Hamiltonians*, Rep. Progr. Phys. **70** (2007), no. 6, 947–1018.

[BH04]   R. C. Brown and D. B. Hinton, *Relative form boundedness and compactness for a second-order differential operator*, J. Comput. Appl. Math. **171** (2004), no. 1-2, 123–140.

[Bog74]   J. Bognár, *Indefinite inner product spaces*, Springer-Verlag, New York, 1974, Ergebnisse der Mathematik und ihrer Grenzgebiete, Band 78.

[BS91]   F. A. Berezin and M. A. Shubin, *The Schrödinger equation*, Mathematics and its Applications (Soviet Series), vol. 66, Kluwer Academic Publishers Group, Dordrecht, 1991, Translated from the 1983 Russian edition by Yu. Rajabov, D. A. Leïtes and N. A. Sakharova and revised by Shubin, With contributions by G. L. Litvinov and Leïtes.

[Cal04]   E. Caliceti, *Real spectra of $\mathscr{PT}$-symmetric operators and perturbation theory*, Czechoslovak J. Phys. **54** (2004), no. 10, 1065–1068.

[Cal05]   _____, *On the spectra of a class of PT-symmetric quantum nonlinear oscillators*, Czechoslovak J. Phys. **55** (2005), no. 9, 1077–1080.

[CCG06]   E. Caliceti, F. Cannata, and S. Graffi, *Perturbation theory of $\mathscr{PT}$ symmetric Hamiltonians*, J. Phys. A **39** (2006), no. 32, 10019–10027.

[CCG08]   _____, *An analytic family of $\mathscr{PT}$-symmetric Hamiltonians with real eigenvalues*, J. Phys. A **41** (2008), no. 24, 244008, 6.

[CG05]   E. Caliceti and S. Graffi, *On a class of non self-adjoint quantum nonlinear oscillators with real spectrum*, J. Nonlinear Math. Phys. **12** (2005), no. suppl. 1, 138–145.

[CG08]   _____, *A criterion for the reality of the spectrum of PT-symmetric Schrödinger operators with complex-valued periodic potentials*, Atti Accad. Naz. Lincei Cl. Sci. Fis. Mat. Natur. Rend. Lincei (9) Mat. Appl. **19** (2008), no. 2, 163–173.

[CGS05]   E. Caliceti, S. Graffi, and J. Sjöstrand, *Spectra of PT-symmetric operators and perturbation theory*, J. Phys. A **38** (2005), no. 1, 185–193.

[Cue]   J.-C. Cuenin, *Spectral inclusions for unbounded block operator matrices with two spectral components and diagonalization of dirac operators*, thesis in preparation.

[DDT01a]   P. Dorey, C. Dunning, and R. Tateo, *Spectral equivalences, Bethe ansatz equations, and reality properties in $\mathcal{PT}$-symmetric quantum mechanics*, J. Phys. A **34** (2001), no. 28, 5679–5704.

[DDT01b]   ———, *Supersymmetry and the spontaneous breakdown of $\mathcal{PT}$ symmetry*, J. Phys. A **34** (2001), no. 28, L391–L400.

[DS88]   N. Dunford and J. T. Schwartz, *Linear operators. Part III*, Wiley Classics Library, John Wiley & Sons Inc., New York, 1988, Spectral operators, With the assistance of William G. Bade and Robert G. Bartle, Reprint of the 1971 original, A Wiley-Interscience Publication.

[EE87]   D. E. Edmunds and W. D. Evans, *Spectral theory and differential operators*, Oxford Mathematical Monographs, The Clarendon Press Oxford University Press, New York, 1987, Oxford Science Publications.

[Fle99]   A. Fleige, *Non-semibounded sesquilinear forms and left-indefinite Sturm-Liouville problems*, Integr. Equ. Oper. Theory **33** (1999), no. 1, 20–33.

[GGK90]   I. Gohberg, S. Goldberg, and M. A. Kaashoek, *Classes of linear operators. Vol. I*, Operator Theory: Advances and Applications, vol. 49, Birkhäuser Verlag, Basel, 1990.

[GMMN09] F. Gesztesy, M. Malamud, M. Mitrea, and S. Naboko, *Generalized polar decompositions for closed operators in Hilbert spaces and some applications*, Integr. Equ. Oper. Theory **64** (2009), no. 1, 83–113.

[Gol66]   S. Goldberg, *Unbounded linear operators: Theory and applications*, McGraw-Hill Book Co., New York, 1966.

[GSZ05]   U. Günther, F. Stefani, and M. Znojil, *MHD α^2-dynamo, Squire equation and $\mathcal{PT}$-symmetric interpolation between square well and harmonic oscillator*, J. Math. Phys. **46** (2005), no. 6, 063504, 22.

[Hes69]   P. Hess, *Zur Störungstheorie linearer Operatoren: Relative Beschränktheit und relative Kompaktheit von Operatoren in Banachräumen*, Comment. Math. Helv. **44** (1969), 245–248.

[Jap02]   G. S. Japaridze, *Space of state vectors in $\mathcal{PT}$-symmetric quantum mechanics*, J. Phys. A **35** (2002), no. 7, 1709–1718.

[Jör67]   K. Jörgens, *Zur Spektraltheorie der Schrödinger-Operatoren*, Math. Z. **96** (1967), 355–372.

[Kat95]   T. Kato, *Perturbation theory for linear operators*, Classics in Mathematics, Springer-Verlag, Berlin, 1995, Reprint of the 1980 edition.

[Lan62]   H. Langer, *Zur Spektraltheorie J-selbstadjungierter Operatoren*, Math. Ann. **146** (1962), 60–85.

[Lan82]   _____, *Spectral functions of definitizable operators in Kreĭn spaces*, Functional analysis (Dubrovnik, 1981), Lecture Notes in Math., vol. 948, Springer, Berlin, 1982, pp. 1–46.

[LT04]   H. Langer and Ch. Tretter, *A Krein space approach to PT-symmetry*, Czechoslovak J. Phys. **54** (2004), no. 10, 1113–1120.

[LT06]   _____, *Corrigendum to: "A Krein space approach to PT symmetry" [Czechoslovak J. Phys. **54** (2004), no. 10, 1113–1120]*, Czechoslovak J. Phys. **56** (2006), no. 9, 1063–1064.

[LTU96]   C.-K. Li, N.-K. Tsing, and F. Uhlig, *Numerical ranges of an operator on an indefinite inner product space*, Electron. J. Linear Algebra **1** (1996), 1–17 (electronic).

[Mos02]   A. Mostafazadeh, *Pseudo-Hermiticity versus PT symmetry: the necessary condition for the reality of the spectrum of a non-Hermitian Hamiltonian*, J. Math. Phys. **43** (2002), no. 1, 205–214.

[MV97]   R. Meise and D. Vogt, *Introduction to functional analysis*, Oxford Graduate Texts in Mathematics, vol. 2, The Clarendon Press Oxford University Press, New York, 1997, Translated from the German by M. S. Ramanujan and revised by the authors.

[Nel64]   E. Nelson, *Interaction of nonrelativistic particles with a quantized scalar field*, J. Math. Phys. **5** (1964), 1190–1197.

[RS75]   M. Reed and B. Simon, *Methods of modern mathematical physics. II. Fourier analysis, self-adjointness*, Academic Press [Harcourt Brace Jovanovich Publishers], New York, 1975.

[RS78]   _____, *Methods of modern mathematical physics. IV. Analysis of operators*, Academic Press [Harcourt Brace Jovanovich Publishers], New York, 1978.

[RS79]   _____, *Methods of modern mathematical physics. III*, Academic Press [Harcourt Brace Jovanovich Publishers], New York, 1979, Scattering theory.

[RS80]   _____, *Methods of modern mathematical physics. I*, second ed., Academic Press Inc. [Harcourt Brace Jovanovich Publishers], New York, 1980, Functional analysis.

[Shi02]   K. C. Shin, *On the reality of the eigenvalues for a class of $\mathcal{PT}$-symmetric oscillators*, Comm. Math. Phys. **229** (2002), no. 3, 543–564.

[Shi04]   _____, *on the shape of spectra for non-self-adjoint periodic Schrödinger operators*, J. Phys. A **37** (2004), no. 34, 8287–8291.

[Shi05]   _____, *Eigenvalues of $\mathcal{PT}$-symmetric oscillators with polynomial potentials*, J. Phys. A **38** (2005), no. 27, 6147–6166.

[Sim70]   B. Simon, *Coupling constant analyticity for the anharmonic oscillator. (With appendix)*, Ann. Physics **58** (1970), 76–136.

[Sim71a]   _____, *Hamiltonians defined as quadratic forms*, Comm. Math. Phys. **21** (1971), 192–210.

[Sim71b]   _____, *Quantum mechanics for Hamiltonians defined as quadratic forms*, Princeton University Press, Princeton, N. J., 1971, Princeton Series in Physics.

[Tan06]   T. Tanaka, *General aspects of $\mathcal{PT}$-symmetric and $\mathcal{P}$-self-adjoint quantum theory in a Krein space*, J. Phys. A **39** (2006), no. 45, 14175–14203.

[Tan07]   _____, *$\mathcal{N}$-fold parasupersymmetry*, Modern Phys. Lett. A **22** (2007), no. 29, 2191–2200.

[Tes09]   G. Teschl, *Mathematical methods in quantum mechanics*, Graduate Studies in Mathematics, vol. 99, Amer. Math. Soc., Providence, RI, 2009, With applications to Schrödinger operators.

[Ves72a]   K. Veselić, *On spectral properties of a class of J-selfadjoint operators. I*, Glasnik Mat. Ser. III **7(27)** (1972), 229–248.

[Ves72b]   _____, *On spectral properties of a class of J-selfadjoint operators. II*, Glasnik Mat. Ser. III **7(27)** (1972), 249–254.

[Ves08]   _____, *Spectral perturbation bounds for selfadjoint operators. I*, Oper. Matrices **2** (2008), no. 3, 307–339.

[Wei00]   J. Weidmann, *Lineare Operatoren in Hilberträumen. Teil 1*, Mathematische Leitfäden. [Mathematical Textbooks], B. G. Teubner, Stuttgart, 2000, Grundlagen. [Foundations].

[Wer07]   D. Werner, *Funktionalanalysis*, sixth ed., Springer-Verlag, Berlin, 2007.